PEOPLE IN THE
COUNTRYSIDE

Tony Champion is Senior Lecturer in geography at the University of Newcastle upon Tyne. His principal research interests lie in the monitoring and analysis of population and social change in Britain and their implications for towns and countryside. He is chair of the IBG Working Party on Migration in Britain (1988–91) and is a past Secretary of the IBG Population Geography Study Group (1987–90). Recent publications include *Counterurbanisation* (Arnold, 1989) and *Contemporary Britain: A Geographical Perspective* (Arnold, 1990). He also contributed to the ACORA report, *Faith in the Countryside* (ACORA and Churchman, 1990).

Charles Watkins is Assistant Director of the Centre for Rural Studies at the Royal Agricultural College, Cirencester. His main research interests include rural geography, land management and landscape history. He is currently researching the legal and clerical professions in rural Britain, the management of woodland and grassland and public controls over land use. Recent books include *Woodland Management and Conservation* (David and Charles, 1990) and he is co-author of *Justice outside the city: access to legal services in rural Britain* (Longman, 1991).

PEOPLE IN THE COUNTRYSIDE

Studies of Social Change in Rural Britain

Edited by
TONY CHAMPION
and CHARLES WATKINS

P·C·P

Paul Chapman
Publishing Ltd

First published 1991 by
Paul Chapman Publishing Ltd
144 Liverpool Road
London
N1 1LA

British Library Cataloguing in Publication Data
People in the countryside: studies of social change in rural Britain.
   rural Britain.
   1. Great Britain. Rural regions. Social conditions
   I. Champion, Tony   II. Watkins, Charles
   941.009734

   ISBN 1-85396-128-0

Typeset by Best-set Typesetters Ltd, Hong Kong
Printed in Great Britain by
Athenæum Press Ltd, Newcastle upon Tyne

C D E F G 7 6

CONTENTS

LIST OF CONTRIBUTORS

Philip Bell is Lecturer in planning at the University of Manchester. He was previously research fellow and lecturer in human geography at St David's University College, Lampeter. His current research interests include rural housing provision. He is co-editor of *Deregulation and Transport* (David Fulton 1990).

Ian Bowler is Senior Lecturer in geography at the University of Leicester. His main research interests are in the geography of agriculture in developed market economies, especially in Western Europe. His books include *Government and Agriculture* (Longman, 1979) and *Agriculture under the C.A.P.* (Manchester University Press, 1985). Currently he is researching rural development processes.

Gordon Clark was an undergraduate and a postgraduate at the University of Edinburgh and he is now a Lecturer in Geography at Lancaster University. His research interests in agriculture include the study of structural change, the agricultural land market and the development of farm management as a career.

Paul Cloke is Professor of human geography at St David's University College, and editor of *Journal of Rural Studies*. His research interests in the role of state policy and planning in the social and economic restructuring of the countryside have most recently been expressed in a book co-authored with Jo Little called *The Rural State?* (OUP, 1990).

Robert Gant is Principal Lecturer in geography at Kingston Polytechnic. His recent publications relate to the personal mobility problems and coping strategies of the rural elderly, local labour markets, and the methodology and practice of oral history.

Sarah Harper is Lecturer in geography at Royal Holloway and Bedford New College, University of London. Her research centres on the social implications of demographic change, with particular emphasis on the rural-urban interface. She has also worked in the UK, USA and China, on the problems faced by societies as they age.

Ann Hockey is a Senior Planning Officer in the Department of Planning and Development Services, Colchester Borough Council. She is currently concerned with developing a computerised information service within the Department's Planning Policy Team. Her research interests include rural policy, population and education.

Gareth Lewis is a Senior Lecturer in geography at the University of Leicester. His current research interests include household decision-making, turnover and population change in rural England.

Jo Little is a Senior Lecturer in Town and Country Planning at Bristol Polytechnic. Her main research interests include rural planning and women's employment. She is co-author of *The Rural State?* (OUP, 1989).

Alison McCleery is a Senior Lecturer in geography in the Business School of Napier Polytechnic, Edinburgh. She is Newsletter Editor for the Population Geography Study Group of the Institute of British Geographers and was a consultant to the Industry Department for Scotland in connection with the 1987 Treasury Review of the Highlands and Islands Development Board. She has written various articles on both population and rural development and has recently been appointed expert consultant to the Population, Employment and Vocational Training Division of the Council of Europe.

Martin Phillips is currently Research Fellow in human geography at St David's University College, Lampeter. He was formerly at the University of London and at Exeter University. Current research interests include historical geography, uneven development, social recomposition and critical theory.

David Rankin was formerly Senior Lecturer in geography at the University of Auckland. He has published widely in the fields of regional planning, labour market segmentation and rural change. He now works as a freelance research consultant based in Norfolk.

Mark Shucksmith is Senior Lecturer and Research Coordinator in the Department of Land Economy, University of Aberdeen. He is currently researching changing attitudes to residential development in the countryside, and farm household strategies and pluriactivity. He is the author of *No Homes For Locals?* (1981), *Rural Housing in Scotland* (1987) and *Housebuilding in Britain's Countryside* (1990).

José Smith is a Senior Lecturer in geography at Kingston Polytechnic. Her research interests include both Third World rural development strategies and transport provision and planning in rural Britain.

Alan Townsend is Reader in geography at the University of Durham and a qualified planner. He specialises both in studying the changing map of UK employment and is Vice Chairman of the (County) Durham Rural Community Council. He brings these two interests together in this volume and in *Contemporary Britain* (Edward Arnold, 1990) with A. G. Champion.

Nigel Walford is a Senior Lecturer in Geographical Information Systems at Kingston Polytechnic. He was formerly Chief Research Officer at the ESRC Data Archive, University of Essex, where his responsibilities included the provision of a user service for government data and the development of the Rural Areas Database. His current research interests include social and economic restructuring in the countryside, the rural environment and the application of geographical information systems to rural databases.

PREFACE

Coping with the urban–rural shift is emerging as the central planning issue of the
1990s, just as the north–south divide dominated discussions of spatial policy in
the 1980s and the inner-city problem took great prominence in the decade before
that. The economic recovery and associated property boom of the late 1980s
led to an acceleration in population movements away from the larger cities
to smaller towns and the countryside. Meanwhile, rural areas are facing
problems in adjusting to the legacy of the last period of strong growth in the
early 1970s – particularly, increasing numbers of elderly people and the job
cutbacks in new industries and branch plants. The countryside is also under
threat from changes in the agricultural regime and from the further centraliza-
tion of services into larger settlements resulting from privatization and other
factors.

Already these changes have brought widespread concern. Most heavily
publicized at the moment is the battle for building sites in the more pressurized
shire counties, where traditional rural interests have combined with incomers
from cities to oppose new development. Increasing attention is, however, being
given to the housing problems of the lower paid and their adult children, as well
as to the lack of access to places with services by large numbers of people
without private transport, including less mobile members of wealthier households. These sorts of questions have led to the recent proliferation of pressure
groups, such as Rural Voice and Acre, as well as prompting special inquiries by
the Town and Country Planning Association and the Church of England.

The principal aim of this book is to increase awareness of the very real
changes occurring in Britain's rural areas and the impact they are having on
people's lives. Partly because of Britain's long-standing anti-urban culture, it is
necessary to challenge the image of the 'rural idyll' of peace and prosperity,
which many of the nation's 80 per cent city dwellers hold. The reality is one of
increasing diversity and conflicting interests, along with a generally much lower
level of access to economic and social opportunities compared to suburbanites of
similar social standing and, in some respects, also in comparison with inner-city
dwellers.

The focus of the book is primarily on the outcome of these conditions rather than on the underlying processes of change themselves: hence the theme of people and social change. The book has two principal topics running through it. First, it documents the changes in the composition of the rural population in terms of age distribution, socioeconomic structure and other social and economic characteristics. Second, it examines the types of problems various groups encounter and the way in which they react to these difficulties with or without the assistance of supporting agencies.

Following Chapter 1's overview of the principal changes taking place in rural Britain, the book selects a number of key issues for more detailed examination. Chapters 2, 3 and 4 deal with aspects of migration and housing, looking at the types and motivations of newcomers to the countryside, the way in which policy can tend to encourage the gentrification of rural areas, and the housing and planning problems the arrival of wealthy people generate for less well-off local people. The next three chapters are concerned with employment trends and related issues, particularly with the major changes that have taken place in the scale and nature of agricultural labour, the extent to which the decline in traditional male manual farmwork has been compensated for by the growth of new forms of employment in rural areas and the question as to whether job opportunities for women there have improved as a result.

The last five chapters raise several important issues. Among them, transport looms large: limited mobility is a major constraint on women seeking work, but takes on even greater significance for the elderly and handicapped. The privatization of 'public transport', while offering more flexible services, has raised fears about a further decline in accessibility for people living in the countryside. The problems of the availability of, and access to, jobs and facilities are particularly acute in a rugged peripheral region such as northern Scotland, but similar difficulties are experienced in maintaining communities in rural England. Agencies such as the Highlands and Islands Development Board and the Rural Development Commission play an important role, but there is a difficult balance to be struck between 'top-down' and 'bottom-up' approaches. Given the diversity of rural Britain, a key challenge in community development projects (besides the need for funds and willing helpers) is accurate and up-to-date information on local conditions.

A key feature of the book is the use of local case-study material to illustrate the major themes identified from a broader evaluation of the topics. A significant proportion of most of the chapters is devoted to a case study undertaken by the authors – in most cases quite specific but in total providing a wide coverage of different local environments across the country (see Figure 1.1, p. 5). This approach is designed to give readers a better feel for the nature and implications of social change in the countryside than can be obtained from standard texts on rural geography. With all the case studies being set in their wider policy and theoretical contexts, the book is intended to be as useful to policy-makers and other professionals involved in rural affairs as to undergraduates studying rural aspects of geography, planning, sociology, economics and environmental sciences.

A subsidiary purpose of the book is to demonstrate the value of geographical perspectives on the countryside. Though geographers have traditionally shown a much stronger interest in urban social change than in rural areas, there is currently an impressive amount of rural research being undertaken by the geographical community. This was evident at the joint meeting of the Rural Geography and Population Geography Study Groups at the Institute of British Geographers 1989 Annual Conference at Coventry Polytechnic: indeed, this provided the chief stimulus for this book. Several of the chapters here are revised versions of the papers presented at that meeting, while others have been specially commissioned subsequently.

In an edited volume such as this, the principal acknowledgements go to the contributors not only for being prepared to participate but also for their forbearance in the face of the editors' foibles: with relatively few murmurings they agreed to see their papers reduced in length to fit the publisher's requirements and, where necessary, to rewrite sections. More specifically we would like to thank Catherine Blishen and Marianne Lagrange of Paul Chapman Publishing for their help and encouragement and Jan MacLaran for typing the bibliography. Finally, it should be noted that several of the authors have made their own acknowledgements at the end of their chapters.

Tony Champion
Newcastle upon Tyne
Charles Watkins
Cirencester
August 1990

1

INTRODUCTION: RECENT DEVELOPMENTS IN THE SOCIAL GEOGRAPHY OF RURAL BRITAIN

Tony Champion and Charles Watkins

The last few years have witnessed an explosion of concern about the ways in which the countryside is changing in Britain and many other countries. Across the European Community, 'Rural society is in a period of upheaval' (Economic and Social Committee, 1989, p. 2), and is '. . . in flux' (European Commission, 1988, p. 23). According to a British group concerned with the future planning of the countryside, 'great changes will face the countryside in the near future' (Town and Country Planning Association 1989, p. 9). The House of Lords' report (1990, p. 7) on the future of rural society recognizes that, while 'historically, rural economies have been based on the exploitation of the natural resources of the countryside . . . , now priorities have changed'. The Archbishops' Commission on Rural Areas (ACORA), reporting in September 1990, also admits the existence of a crisis in the countryside.

This burgeoning of interest seems to revolve around two main sets of issues. One relates to resource use and management and to issues about the conservation of the natural environment and landscape, while the other concerns the welfare of people living and working in the countryside. The two are to some extent interlinked but it is the second of these that constitutes the primary topic of this book. Paralleling the rise of public concern in rural people and their well-being, there has been a very considerable growth in the scale of research in the area of rural social geography, particularly its more applied aspects (Cloke, 1990; Cloke and Moseley, 1990). Nevertheless, even in 1990, it was possible for the House of Lords (1990, p. 47) to comment: 'To define rural areas is not easy . . . , it is even more difficult to discover much about the people who live there'. This book attempts to rectify this deficiency in some measure. The first purpose of this chapter is to provide a fuller justification of the book in terms of its place within the evolving literature and of the way it approaches its task. It then goes on to outline the principal features of people in the countryside and the problems they face, in order to provide a context for the more specific studies presented in subsequent chapters.

Studies of people in the countryside

There has long been an interest in the study of social change in the countryside. Thomas, writing in 1939, considered that (p. 179) 'books concerned with technical and social problems of the countryside produced some of the best writing of the first thirty years of this century'. During the Second World War, a classic study of the needs for rural reconstruction was edited by Orwin (1944) and, during the 1950s, a number of important social studies of rural life were undertaken (Williams, 1963). For much of this century the driving force behind many studies of rural social change was that of depopulation (Saville, 1957; Bracey, 1958; Lawton, 1968). It was not until the early 1970s that evidence for a turnaround in population trends began to emerge (Hall *et al.*, 1973; Lawton, 1977; Spence *et al.*, 1982) and this was amply confirmed when the results of the 1981 Census became available (Randolph and Robert, 1981; Champion, 1982; Hamnett and Randolph, 1983). The growth in the rural population has been paralleled by a growing interest in rural social change (Frankenberg, 1966; Newby *et al.*, 1978; Phillips and Williams, 1984). Some writers applied sociological concepts, such as social class and conflict, to their analysis of rural society, challenging the established notions of rural harmony and consensus (Pahl, 1966; Ambrose, 1974). Newby (1979) showed how the arrival of former urban dwellers in search of a mythical 'rural idyll' was removing what he called the 'occupational community' based on agriculture, and was creating social divisions that had the potential for undermining the 'social fabric of rural community life'. There was also concern about the effect that population changes were having on the provision of services and facilities in the countryside – in particular, the comparative disadvantage suffered by specific social groups, such as the elderly, the disabled and mothers with young children (Moseley *et al.*, 1977; Moseley, 1979; Shaw, 1979; McLaughlin, 1986).

Geographical research has played an increasing part in the growth and re-orientation of rural studies, particularly over the last ten years. Indeed, Cloke and Moseley (1990) emphasize the strength of this growth by pointing to the lack of geographical interest in rural change twenty years ago and tracing the origins of 'rural geography' as a recognized sub-discipline in Britain to the early 1970s. This was evidenced by the publication of Hugh Clout's (1972) *Rural Geography: An Introductory Survey* and by the setting up by the Institute of British Geographers of the Rural Geography Study Group in 1974. In replacing a study group devoted to agricultural geography, the latter marked the end of a period dominated by research on the farming sector and the opening up of much broader interest, as reflected most notably in the appearance of two major journals – the *Countryside Planning Yearbook* published annually since 1980 and relaunched in 1988 as the *International Yearbook of Rural Planning*, and the quarterly *Journal of Rural Studies* launched in 1985. Through an analysis of research registers, Clark (1982a) was able to show that, by the early 1980s, the interests of British rural geographers had moved significantly away from agriculture, land use and settlement with increased attention to other rural resources, population, transport and rural development and planning. Since then, this tendency appears to have gained further momentum; only 2 of 16 'important texts used in rural geography' and published in the 1980s dealt exclusively with agricultural matters, while the other topics included employment, housing, accessibility, rural settlements, countryside planning, land use and recreation (Cloke and Moseley, 1990, p. 120).

Rural geography has also seen substantial conceptual and methodological development in the last few years. In the past, there has been a concern that most of the new thinking in social geography was coming from urban geography – to the extent that, according to Phillips and Williams (1984), some saw social geography as being synonymous with urban social geography. Cloke (1980) complained of the 'conceptual famine' of rural geography, while Moseley (1980a) considered that rural geography was lagging behind in its methodology. Since then a small but growing band of rural geographers led by Cloke (1990) has championed the cause of greater conceptual sophistication, arguing not only for greater attention to be given to the role of institutional constraints but also for a move towards the better understanding of how planning and policy-making are tied up with the wider role of the State and, in turn, with the power relations that underlie the State (see also Cloke and Little, 1990).

In this context, the present book has several purposes. First and foremost, it constitutes a further contribution to the study of social change in the countryside and, in collecting together a number of research studies, emphasizes the strength of British rural geography. Second, while the authors each follow up their specific topics, their contributions are informed by a holistic view, reinforcing the broader perspective that is now being taken on the changing countryside. Third, the essays taken together bear witness to the rich variety of conceptual approaches that are now being employed in rural geography, ranging from pure description and monitoring of population and employment trends through the behavioural aspects of the decision to move into the countryside to the examination of the ways in which institutional factors affect people's opportunities and activities and how these factors are themselves moulded by wider political and social contexts. Finally, and partly related to this, the chapters in this book reflect the importance attached to current issues and policy relevance by rural geography in Britain – a powerful element in its growth over the last twenty years.

A central feature of this book, as mentioned in the Preface, is the presentation of original case-study material in most of the chapters. To some extent, this is a consciously pragmatic decision in that this book is designed to complement more broadly based texts on the countryside (e.g. Pacione, 1984a; Phillips and Williams, 1984; Gilg, 1985; Robinson, 1990), which cannot spare the space for the full treatment of case studies. It is also true that, as pointed out in Chapter 12, standard data sources have not been able to keep up with the pace of change in the types of questions being asked about people in the countryside, with the result that researchers have to collect their own data. At the same time, however, the emphasis on case studies reflects the fact that these questions are now much more related to the processes of change rather than to the resulting patterns, and can be answered satisfactorily only by recourse to in-depth studies, often restricted to a very specific geographical context.

Finally, mention should be made of the book's primary concern with Britain. Moseley, in Cloke and Moseley (1990, p. 125), has commented on the 'excessively parochial' nature of British rural geography, rarely straying outside the Anglo-American arena. In this sense, the book is therefore reflecting the current nature of the subject. Taken together, the case studies documented in subsequent chapters of this book provide an insight into the range of conditions found across rural Britain (see Figure 1.1). On the other hand, the book is not designed to provide an inventory of rural Britain but rather to shed light on the scale and nature of changes that have taken place there. As such, many of the underlying process and the issues arising from them have their counterparts in other countries and can thus be taken forward as the subject of comparative study (see Lowe and Bodiguel, 1990; Robinson, 1990). The policy implications, however, need to be handled very carefully, not only because of the differences *between* countries in institutional arrangements, customs, and so forth, but also because of the wide variety of rural contexts *within* even a single country like Britain.

Definitions of the countryside

Thus far in this chapter we have used the terms 'rural' and 'countryside' interchangeably and, by referring to a substantial and growing body of literature on people in the countryside, we have given the impression that there exists a clearly defined and widely accepted subject area. In one sense this is true, but a number of critics have argued cogently against drawing a distinction between urban and rural locales. The issues need to be rehearsed briefly here in order to justify – or excuse – the essentially pragmatic approach adopted in this book and to emphasize the need for caution in interpreting statistics that purport to relate to 'rural Britain' or some similar description.

Judging from the previous literature, the central problem arises from the progressive erosion of urban–rural distinctions (see Phillips and Williams, 1984). In the days of walled cities, it was relatively easy to distinguish town from country and, indeed, well into the twentieth century it was considered by some that urban areas were distinctive by virtue of their population size, density and heterogeneity. Support for the idea of a rural–urban dichotomy has, however, fallen away in the face of the outward spread of urban influences into the countryside. As an alternative, some sociologists, such as Frankenberg (1966), proposed a rural–urban continuum along which most communities can be arranged. Then again others, such as Pahl (1968), were arguing that all of Britain was now culturally urbanized and viewed lifestyle as being more dependent on class and the stage in the family cycle than on a particular location.

Support for this view has grown stronger with the onset of counterurbanization. The arrival of ex-urbanites in some of the physically more remote parts of

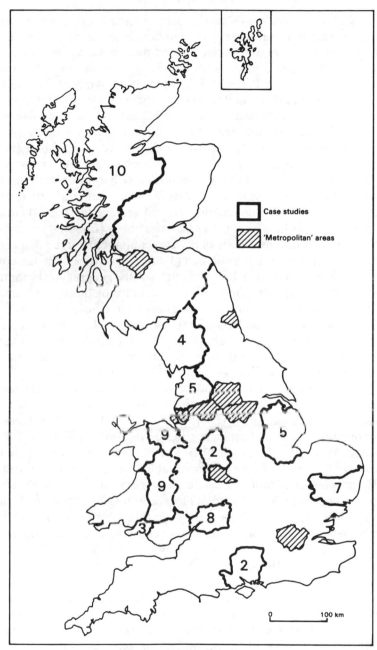

Figure 1.1   Location of case studies in this book. The map shows counties of England and Wales containing case studies presented in this book, together with the (pre-1986) area of the Highlands and Islands Development Board. Numbers refer to the chapters in which the case studies appear.

the countryside has led towards greater socioeconomic and cultural heterogeneity there as well as in rural areas located closer to the major urban centres. Various classifications of the social groups that make up the new types of communities have been made over the years, indicating the variety of newcomers (Stamp, 1949; Pahl, 1966; Thorns, 1968; Ambrose, 1974). Cullingford and Openshaw (1982) found substantial socioeconomic differences existing in the 'rural' areas of north-east England, and Dunn, Rawson and Rogers (1981) were able to identify six different types of community on the basis of their 'housing profiles' in an area as small as South Oxfordshire District (outside its towns). In this book, Sarah Harper is able to demonstrate the variety of motives that causes people to move into the countryside (Chapter 2) while, in terms of employment structure, Alan Townsend reveals the rather small contrast that now exists between rural areas and the national average, particularly in terms of the proportion of workers engaged in manufacturing (Chapter 6).

Nevertheless, a substantial body of opinion continues to argue that rural areas retain distinctive features. Even Hoggart (1988), whose assault on the concept of 'rural' is the most persuasive to be launched from the geographical community to date, is forced to recognize the continuing validity of three of Clout's (1984) attributes of rural areas – namely, a low population density, loose networks of infrastructure and services and a landscape dominated by farming and forestry. Hoggart goes on to deride the result, saying that 'the sum effect is to suggest that "rural" areas contain small settlements separated by open countryside. This is usually the understanding of the general public' (Hoggart, 1988, pp. 35–6). This latter point, however, is not so lightly dismissed by others, particularly those who feel that the countryside is given a basic distinctiveness by its inaccessibility and isolation. For instance, Moseley (1980b) suggests that, while urban and rural areas share a set of common problems, there are also problems specific to both types of area stemming from their place-specific features – in rural areas, these being essentially those of access, demographic imbalance and infrastructural costs. Nationwide classifications that include rural districts always discover a clear rural component, whether they include an explicit 'distance from urban centre' measure, such as the 'index of rurality' of Cloke (1977) and Cloke and Edwards (1986), or focus entirely on socio-demographic characteristics, such as Webber and Craig (1977) and Dunn, Rawson and Rogers (1981). Another traditional feature of the countryside, which remains fairly distinctive and indeed has been reinforced by recent developments, is the fact that rich and poor often live side by side. This is increasingly a source of friction magnified by the control the rich are generally able to exercise over the allocation of goods and services that affect their neighbours. Indeed, Phillips and Williams (1984) place the rural–political framework alongside the distance-accessibility factor, maintaining that these are the two principal features that distinguish rural from urban areas.

Clearly, the whole thrust of this book is that 'people in the countryside' do constitute a distinctive topic. This approach can be justified in terms of the argument just outlined, namely, that rural areas continue to have some particular characteristics that set them apart from urban areas. The criticisms of

Hoggart (1988) are deemed to carry less weight in the present context because, whereas he was assessing the validity of 'rural' as an explanatory tool, this book focuses primarily on the outcomes of recent developments in the form of their effects on people, their characteristics and their problems. This said, however, what we are not going to do is to argue the case for one particular definition of rural Britain, preferring to stay with 'the understanding of the general public', as Hoggart put it – that is, areas with 'small settlements separated by open countryside'. This provides the palimpsest within which the studies presented in subsequent chapters are set. Unfortunately, this approach does cause problems of comparability, particularly where statistics are used that purport to relate to some definition of 'rural Britain'. Alan Townsend's analysis of employment trends (Chapter 6) illustrates this point very nicely since this explicitly recognized three definitions of rural areas – rural counties based on overall population density necessarily containing quite large towns, rural districts as defined by Webber and Craig (1977) and generally excluding larger settlements, and Rural Development Areas, which are groupings of parishes and represent the 'deep countryside' with its abundant problems of inaccessibility and isolation. For each chapter in this book, the authors have chosen the frames of reference they feel are most appropriate, so readers are urged to exercise care in interpreting statistics relating to these areas.

Population change

It is the resurgence of population growth in rural areas that constitutes the key symptom, if not the cause, of the fundamental changes that are now occurring in the countryside. As mentioned above, even twenty five years ago discussion of rural affairs was dominated by the issue of depopulation. Though some statistics showed that rural Britain was experiencing population growth at the time, this was really a quirk of the statistical areas used – the so-called 'Rural Districts' of the pre-1974 local government system, the boundaries of which were so out of date that these areas were accommodating a substantial part of the country's urban and suburban growth (Lawton, 1977). When the rural population turnaround was first identified in the 1970s, a central research issue was whether this, too, resulted from the underbounding of urban areas and merely constituted a continuation of local metropolitan decentralization or whether it represented a 'clean break' from the past and a move towards some kind of new 'post-industrial' settlement pattern (see, for instance, Hamnett and Randolph, 1983). This debate has not been confined to Britain, but was initiated in the USA (Morrison and Wheeler, 1976; Vining and Strauss, 1977; Gordon, 1979) and has subsequently been addressed in most countries of the Developed World (for a review and national case studies, see Champion, 1989a). In trying to pin down the nature of this national and international phenomenon, particular attention has been given to the definitions of the areas being used to analyse population change and to the monitoring of population trends over time.

The statistical evidence marshalled for the UK points to a definite change in population trends affecting extensive rural zones, and this is not just limited to the fringes of large cities. An analysis based on the Centre for Urban and Regional Development Studies (CURDS) Local Labour Market Areas has demonstrated the large scale of the turnaround in the most rural parts of Britain, which had a population growth rate 8.9 percentage points above the national average between the 1971 and 1981 Censuses, compared with one 5.5 points below the average in 1951–61 (Champion, 1989b). Similarly, the 'remoter, mainly rural' districts identified by Webber and Craig (1977) were the only district type recognized by the OPCS (1981b) to have increased their rate of growth between the 1960s and 1970s, moving against the national trend, and, indeed, between 1971 and 1981 this district type recorded faster growth than all the others bar districts containing New Towns. More detailed analysis confirmed that this strong performance by rural areas was not confined to more accessible places or a particular part of Britain but was extremely widespread, with some of the largest upward movements in change rate occurring in the most rural and remote areas (Champion, 1982). These results were confirmed by more localized studies of Cornwall (Perry, Dean and Brown, 1986), North Devon (Bolton and Chalkley, 1990) and the Scottish Highlands (Jones *et al.*, 1986).

The more detailed monitoring of trends over time has sought to establish whether the turnaround is a long-term feature likely to dominate rural affairs for the foreseeable future or whether it resulted from a chance combination of factors, which may not be repeated. This question was raised by the observation that the rate of rural population growth – using any definition of rural areas – rose swiftly to a peak in the early 1970s and then fell away steadily during the remainder of that decade (Champion, 1981). Subsequent monitoring, however, has revealed a renewed acceleration of population growth and associated migration gains in rural Britain. For instance, while the CURDS Rural Areas saw their rate of growth slip from an annual average of 9.4 per thousand in 1971–81 to 6.3 in 1981–5, the rate was back up to 9.0 in 1985–7. Meanwhile, the 'remoter, mainly rural' OPCS district type averaged an 11.4 per thousand increase between 1984 and 1988, a remarkable recovery from 5.9 in 1981–4 and getting back close to the peak level of 14.5 experienced in the early 1970s (Jones and Armitage, 1990). Indeed, in terms of net migration alone, there was remarkably little difference between the peak level of 13.4 per thousand in 1971–4 and the 11.9 rate for the most recent period. The conclusion to be drawn from these observations is that, rather than the 1970s turnaround being a temporary observation, rural areas have experienced a major long-term transformation into population growth that will merely be affected by cyclic factors associated particularly with the state of the national economy – as was the case during the recession period of the late 1970s and early 1980s.

This bullish interpretation must, however, be qualified in at least two respects. In the first place, the average rates just presented conceal a great variety of local experiences and it would be unrealistic to expect that all parts of rural Britain would have experienced similarly strong growth in the past few years or should expect to do so in the future. This is true even at the broad regional scale where, in the 1980s, the differential in rural growth rates between the northern and southern halves of Britain widened markedly after a decade of convergence – an annual average of 4.7 per thousand in 1981–7 for the CURDS Rural Areas in the North, compared with 10.8 in the South. It is even clearer at more localized scales where, as Cloke (1983b) has argued, particular local circumstances modify the impact of counterurbanization forces. This has been demonstrated by Weekley's (1988) study in the East Midlands where, despite strong overall population growth, a significant number of parishes were experiencing depopulation. Similarly, examination of the 1981 Census parish results for Northumberland reveals that 87 of Northumberland's 157 parishes (55 per cent) lost population between 1971 and 1981, while even in a pressurized county such as West Sussex the proportion was as high as 46 per cent (Champion and Townsend, 1990a). This pattern of localized depopulation results from a variety of factors including the transfer of housing stock into second homes and holiday lets, reduction in household size, planning embargoes on the construction of new homes and the associated policy of concentrating new development and service provision in 'key settlements'.

The second qualification relates to the different experience of population sub-groups. The overall statistics on population growth and net inmigration mask a variety of more complex changes taking place both within rural areas and between countryside and town. Net migration gains merely represent a positive balance between usually much larger flows of inmigrants and outmigrants, with these two groups often contrasting significantly in their composition. This is particularly true of rural Britain today because, despite overall growth, most rural areas are continuing to lose school-leavers and young adults. Of the 17 English counties with the highest rates of net migration gain in 1987, all but two recorded a *net loss* of 16–19-year-olds (Champion and Townsend, 1990a). Generally, though one of the distinctive features associated with the turnaround is the growth of the working-age population, the recent migration gains in rural areas have been disproportionately weighted towards older families and people nearing retirement age, as well as pensioners themselves (see, for instance, Dean *et al*, 1984). Similarly, the basic population counts hide the fact that the newcomers tend to be wealthier and of higher socioeconomic status than local residents (Bolton and Chalkley, 1990), in many cases forcing the latter (and, more particularly, their children) to seek accommodation in the larger rural settlements or leave the countryside altogether. These processes of demographic and social change underlie many of the key issues currently being faced in rural Britain and therefore form a central theme of this book.

Employment

Given that only a relatively small proportion of recent population growth in most of rural Britain can be accounted for by retirement migration and also that only certain parts are readily accessible to larger urban centres for commuting, it is not surprising that recent trends in population have been closely paralleled by overall patterns of employment change. Indeed, the variability in population growth rates between rural areas can also be explained largely in terms of their different employment fortunes. Traditional rural activities, such as farming and forestry, have generally experienced a steady contraction in job numbers, whereas there has been considerable growth in employment in the manufacturing, tourism-related industries and private-sector services. Though the individual growth industries each tend to have more localized distributions than the extensive land-using sectors, both at a local scale and across the country as a whole, together they have brought economic growth to most areas. Some places, however, have not benefited to the same extent and have thus been dominated by the contraction of traditional industries and by associated depopulation.

Major changes have been taking place in employment on farms – changes that would be considered even more dramatic if their general effects were not merely seen as a continuation of the long-term trend in the number and type of jobs. The limited demand for food, the nature of price support and other government assistance, and the need for farmers to adopt new cost-reducing technologies in order to remain economically viable have stimulated the industrialization of agriculture and its production methods (Bowler, 1985). This has involved the specialization of production on farms and in particular farming regions, an increase in the number of larger farms and associated decline in the number of farm businesses and a massive fall in the size of the agricultural workforce, which has halved since 1950 (Munton, Marsden and Whatmore, 1990). The effects on the people working in farming are described in this book by Gordon Clark (Chapter 5), who emphasizes the differences in fortunes between the hired workforce and the farmers themselves, and also between the rapid decline in traditional full-time male jobs and the growth of seasonal, temporary and other forms of casual work, which involve women to a great extent. Perhaps the clearest social consequence of recent trends, however, has been the increasing differentiation among the farmers themselves: on the one hand, a relatively small élite who are closely tied into, and benefit strongly from, the present system (and are the ones most likely to employ farm managers and professional consultants) and, on the other, a larger number of farmers who are forced to compete on increasingly unfavourable terms or are severely handicapped by their limited scale of operations and their inability to sustain a livelihood for their families without engaging in some form of part-time (usually off-farm) activity.

Alongside the industrialization of agriculture has come the 'ruralization of industry' (Healey and Ilbery, 1985). In contrast to the trends in farming, this development appears very remarkable because it was essentially unexpected and flies in the face of long-term trends. Though past depopulation has usually been linked most closely to the contraction of agricultural employment, the decline of traditional craft manufacturing and the concentration of consumer services in larger centres also played a major part in undermining the economy and social basis of the countryside. As outlined by Alan Townsend in this book (Chapter 6), the 'urban–rural' shift in the location of manufacturing has been the most potent element in the rural population turnaround of the 1960s and 1970s – to the extent that today the proportion of the workforce engaged in manufacturing in rural Britain is only marginally lower than the national average. There has also been strong growth in employment in the service sector, a great deal admittedly in public-sector jobs and consumer services ten to twenty years ago but, in the 1980s, much more associated with activities that, like most of manufacturing, are 'export-orientated', catering for more than local demand. Most notable among them are 'tourism-related industries', which recorded a 17-per-cent increase in jobs in just six years (1981–7) in Britain's 'remoter, mainly rural' districts, but they also include banking, insurance, finance and other business services, which grew by 32 per cent over the same period, albeit from a below-average base.

These new sources of growth have, however, by no means solved the rural employment problems caused by the contraction of traditional industries. In the first place, not surprisingly, they have not occurred evenly across the countryside, each being subject to its own locational requirements. One particularly localized development can be seen in the North Sea oil industry, which has strongly affected the Scottish Highlands and Islands, as documented by Alison McCleery in Chapter 10. This was, however, rather unusual among the new activities not only in its very period-specific occurrence but also in its preference for full-time male workers. The latter is noteworthy because, second among the continuing problems of rural areas, is the fact that the majority of new jobs have been in the form of part-time employment, particularly for women. This means that the total amount of work available in rural areas has not increased as fast as the statistics on overall job numbers would suggest. Moreover, as detailed by Jo Little in Chapter 7, much of this work is poorly paid, lacks security and long-term promotion prospects and is carried out in unattractive working conditions, reflecting the lack of unionization and other forms of workers' organization in the countryside. As such it has helped to reinforce the bimodal social-class structure of rural areas, with its dual emphasis on managerial personnel and on semi-skilled manual workers (Errington, 1990).

Policies aimed at tackling employment problems in rural areas have contributed to the development of the new sources of job growth. As outlined in Chapter 11, such bodies as the Rural Development Commission and English Estates have, along with local authorities, placed great emphasis on the development of industrial estates. This has been very useful in supporting the spread of manufacturing industry into rural areas (Shaw and Williams, 1985; Bolton and Chalkley, 1990), particularly as it reinforces one of the main advantages that rural areas possess over the cities – namely, cheaper, more readily available sites (Fothergill *et al.*, 1985). Nevertheless, the achievements of these and other initiatives have been limited. As Ian Bowler and Gareth Lewis point out in Chapter 11, they have tended to bring greater benefits to country towns than to smaller rural settlements: only relatively recently has the need for a greater range of schemes and the merits of a more community-orientated approach been widely appreciated. Many of the new opportunities have been particularly inaccessible to people without private means of transport, including large numbers of married women and young single people. As Jo Little demonstrates in Chapter 7, few local authorities have so far put forward within Rural Development Programmes a policy objective specifically on women's employment issues, let alone have advanced an initiative to tackle their

problems. Policy-making still needs to give greater recognition to the distinctiveness of the geographical context of rural areas as well as the variety of local
conditions found there.

Housing

Rural housing continues to be one of the most fruitful areas of rural research.
This reflects the importance of housing within society as a whole. As Phillips
and Williams (1984, p. 97) have pointed out, 'Houses ... confer upon their
inhabitants much more than merely roofs over their heads; they involve status
and an expression of place in society'. In other words, they are not only places in
which to live but are also important forms of investment and one of the most
tangible forms of positional good (Hirsch, 1978). In Chapter 2, Sarah Harper
looks in detail at the wide range of factors that affect people's decision to move
to the countryside. The availability of different types of housing to some extent
determines the types of people who can afford to live in the countryside. In
many rural parishes there has been little new housing development because of
various planning policies that have encouraged new developments only in
certain selected villages and small towns (Cloke, 1979). Such planning policies
are inflating rural property prices relative to urban ones, which means that it is
only the relatively wealthy who can afford to move into the countryside. Apart
from limited infill and, especially during the 1980s, the conversion of old
agricultural buildings into housing, there has been very little new housing
development. At the same time, many redundant pairs of agricultural workers
cottages have been combined to form single houses of high capital value.
Moreover, the new houses that have been built have usually been large, if not
luxurious, and highly priced. In Chapter 3, Paul Cloke, Martin Phillips and
David Rankin investigate the way in which different elements of the middle
classes gain entry to the housing market in their Gower study area.

The conversion of barns, cow sheds and stables into housing is one of the
most visible indicators of the demand for rural housing. Converted agricultural
buildings provide ideal copy for the enthusiatic estate agent. They usually have a
period feel even if any historical authenticity has been lost in the process of
conversion. Some commentators in the early 1980s saw such conversions as a
way of providing relatively cheap housing for local people. In practice, however,
this was not the case. A study of conversions in West Devon, Boothferry and the
Cotswolds showed that they only rarely provide a solution to rural housing
problems (Watkins and Winter, 1988). It also indicated that there were strong
regional variations in the pressure to convert agricultural buildings. Over the
ten-year period 1977–87, the number of conversions in the Cotswolds was high
but fairly static; in West Devon there was a substantial increase year by year in
the number of applications; while in Boothferry there was little pressure for
conversion at all. These variations were due to a complex mixture of factors
including the physical supply of suitable buildings, and variations in development pressures, local planning-authority policies and property values.

Some of the most useful data on rural housing are provided by McLaughlin
(1986). Data collected in the early 1980s from his five study areas show that both

housing standards and access to housing are important issues. The amount of poor-quality housing varied dramatically between the study areas with two (Suffolk and Shropshire) having twice the national average level of poor housing. By contrast, in Essex and Northumberland where there was a very high proportion of owner-occupation, there was almost no sub-standard housing. The rural areas had a very high proportion of owner-occupied property that was fully owned (43 per cent) compared with the national average (23 per cent). This reflected the high proportion of retired people living in the study areas. In contrast, the proportion of public rented housing (15 per cent) was less than half the national average (33 per cent) at that time. This council housing was unequally distributed with many parishes having none at all. The low proportion of people living in council housing in rural areas is confirmed by data collected in 1989, which showed that only 12 per cent of a random survey of people living in twenty rural parishes lived in public rented accommodation (Davies *et al.*, 1990a). In general, the stock of council housing has diminished over the last ten years with central government's encouragement of council-house sales and discouragement of the building of new council houses. Another problem with this public stock identified by McLaughlin (1986) was that much was under-occupied, with a high proportion of small, elderly households.

Although various efforts have been made to establish housing associations in rural areas, the evidence indicates that only very small numbers of dwellings are concerned and that developments tend not to house the same groups of people who have traditionally relied on council housing. A study of housing associations in Devon, for example, showed that this sector contained a higher proportion of professionals and managers, and a lower proportion of skilled and unskilled workers than did council housing (Richmond, 1987). The amount of private rented and tied housing in rural areas should not be under-estimated. McLaughlin (1986) found that private rented housing was most common, as might be expected, in areas where agriculture and tourism were important employers. Over a fifth (21 per cent) of accommodation was either private rented (13 per cent) or rent-free (8 per cent). Most of the rent-free accommodation was tied to particular jobs and much was of low standard. Two thirds of the inhabitants were in manual occupations. In contrast, almost a half (47 per cent) of those living in private rented housing were in non-manual occupations. This was partly explained by the large numbers of tenant farms and farm managers, but this group also included high-income non-manual employees who could not buy houses because of cost and who were ineligible for council housing. Bowler and Lewis (1987) have shown how important private rented accommodation can be in some traditional estate villages where most or, indeed, in some cases all, the housing is still owned by a single landowner. Their research in Northamptonshire showed that, of the 247 parishes in their study, 29 parishes still had more than 60 per cent of their households in the private rented sector in 1981. Their data also show, however, the extent of the recent decline in this housing sector: in the twenty years from 1961 to 1981, the number of parishes with this high proportion of private rented accommodation more than halved, falling from 74 to 29.

Recent research on rural housing has considered the search for mechanisms to release land for affordable housing using such means as Section 52 planning arrangements or legal covenants (Rogers and Winter, 1988; Clark, 1990; RICS, 1990). Mark Shucksmith investigates various aspects of this question in Chapter 4. A particular problem is the development of suitable selection procedures for prospective tenants and owners of 'social' housing. At present, although various schemes to provide relatively low-cost housing in rural areas have received considerable publicity, few such houses have been built and the future for cheap housing in the countryside appears to be bleak.

Social change and deprivation

It is difficult for anyone to take in the full range of social changes that have taken place in rural areas over the last sixty years. The changes are complex and vary from region to region. A series of very broad, interlinked, social changes, however, can be recognized and the roots of many of these changes can be traced back to well before the Second World War. First, the massive decline in the number of jobs associated with agriculture and with private service has meant that there has been an associated decline in the number of working-class residents. Second, there has been a substantial increase in the number of middle-class residents. Third, there has been (notwithstanding recent changes under the Thatcher administration examined by Philip Bell and Paul Cloke in Chapter 9) a large increase in public expenditure (Bell and Cloke, 1989). Fourth, there has been an astonishing increase in car ownership and improvement in roads, which has had a remarkable effect on the distances people are prepared to travel for work and leisure. Fifth, there has been a dramatic improvement in the quality of housing in rural areas. Finally, the spread of televisions, telephones and recent developments in telecommunications have considerably reduced the isolation of rural areas (Newby, 1990, Watkins, 1990).

Taken together, these changes might suggest that rural society is becoming more homogeneous. Although this is true to some extent, there are many exceptions. In Chapters 2 and 3, for example, it is shown that although many of the people who move to the countryside might be loosely termed 'middle class', in fact there is a complicated range of different groups involved who are moving to the countryside for a wide variety of reasons. It is important to remember in this context that the desire of members of the middle class to live in the countryside has a very long history (Williams, 1973). Until fairly recently, one of the main reasons for moving to the country, whether one was an impoverished poet or a retired army officer, was because it was cheaper to live there than in a town. Thomas (1939, p. 14) noted rather cynically that 'only retired people with small pensions move into the village, not to live, but to die there as slowly as possible'. With improvements in transport and other communications and services during the 1950s and 1960s, rural property began to become more expensive, and now in many areas it is more expensive to purchase property in the countryside than in neighbouring towns. No one now moves to the country in order to save money, unless it be to move into a permanent caravan or make use of relatively cheap winter-lets of holiday accommodation.

One aspect of social change that has captured many people's imagination is the notion of the contrast, and to some extent conflict, between 'incomers' and 'locals' within rural areas. The tension between these two groups informs many of the plot-lines of *The Archers* – the long-running British radio soap opera dealing with country life. It also forms the basis of many journalistic pieces dealing with the conflict between newcomers to an area and existing farming practices, such as the keeping of a large dairy herd on a farm whose buildings are located within a village. There is clearly a good deal of truth in the caricature of the incomer to an area who wants to keep things as they are and protect their idealized view of what constitutes the countryside (Newby, 1990). However, it is important to realize that there is no simple division between incomer and local. Under what circumstances, for example, do the children of incomers become locals? Moreover, incomers, as shown in Chapter 2, may move to an area in a number of stages, buying a cottage that is first used for weekends only, and then moved into permanently in retirement. In this sort of case, the individuals concerned, although moving permanently to an area only recently, may have developed a network of relationships with the people living locally stretching back twenty or thirty years.

Another point is that that part of the population, which might be termed 'local', is not unchanging. It is still common for the children of 'local' families to move away from rural areas, because of relatively limited employment opportunities. Those whose parents are owner-occupiers may be able eventually to move back to their 'home' area, while others will remain away. One group of people usually classed as local is the farming family, but this is not necessarily the case at all. Shawyer's (1990) study of farm family change in eastern Nottinghamshire has shown the complicated nature of family succession to farms and the extent to which the children of farmers choose not to carry on working the family farm, thus allowing change of ownership to take place.

Many of the social changes affecting rural areas have resulted in an improved standard of living for rural residents, but the overall impression of wealth that can be gained from a quick visit to the country is misleading. Many poor people still live in rural areas, and the rural-deprivation debate is still alive. The most significant recent study of rural deprivation in England (although the fieldwork was carried out almost ten years ago) was that carried out by McLaughlin (1986). A total of 876 households were interviewed, representing around 2,400 individuals. A supplementary-benefit entitlement income was calculated for each household and this was used in conjunction with the household's actual gross disposable income to assess whether the household was living in or on the margins of poverty. The survey showed that around a quarter of all households in the five rural areas were living in or on the margins of poverty. The large proportion of households consisting of single elderly people means that around a fifth of the rural population were living in or on the margins of poverty. This large minority of poor people needs to be set against the relatively high proportion of wealthy people in these same five study areas. Townsend (1979) had shown that 11 per cent of households nationally had gross disposable incomes at least three times their supplementary-benefit entitlement income.

The equivalent figure for the five rural study areas was as high as 26 per cent. Moreover, the highest incomes for the rural areas were considerably higher (33 per cent) than the national equivalent and the lowest incomes were comparatively lower. The survey showed that most of the poor living in rural areas were elderly households with State pensions as their sole source of income. Apart from these households, the most important cause of poverty was very low wage levels: almost a quarter (24 per cent) of all male manual workers and over three quarters of female manual workers earned below the 1981 low-pay threshold of £80 per week. The equivalent figures for Great Britain as a whole were 10 and 66 per cent respectively.

The ageing of the rural population is another factor that has great relevance to the deprivation debate. At the national level, population estimates show that the proportion of elderly people will continue to rise into the next century. Moreover, long-term illnesses, and serious mental and physical disability, are becoming compressed into short and later stages of life. With the widespread decline in the provision of community services in rural areas there are indications that there has been a deterioration in the welfare of the rural elderly, especially the disabled (see Chapter 8). In many rural areas, the proportion of elderly is above the national average. The principal cause of such ageing is the in-movement of the retired or prospective retired who choose to live in pleasant rural surroundings and can afford to do so. At the same time, younger members of the established community are leaving these same areas to obtain jobs elsewhere.

Bradley (1987, p. 170), who worked on the McLaughlin project, notes that one of the characteristics of rural as opposed to urban poverty is that 'some of the poorest citizens live cheek-by-jowl with the wealthiest pillars of the community. Looked at from a perspective of relative poverty, the gulfs that exist between the life chances of neighbours are breathtaking'. McLaughlin (1987) suggests that many of the causes of deprivation in rural areas are the result of decisions taken by better-off neighbours, whether acting in a capacity of employer, landlord or political representative. There is, of course, nothing new about the rich and the poor living close together in the countryside. Indeed, the disparity between the two groups was much greater in the late nineteenth century. What is different today is that the poorer groups are often outnumbered by the relatively wealthy and their plight is frequently forgotten. The image of rural well-being is so strong that few people imagine that rural deprivation can still exist. The small amount of detailed research evidence available shows that this is not the case. However, much of this research is dated and further research is required to assess the current extent and nature of rural deprivation.

Service provision

Changes in the extent and character of the provision of various services in rural areas have long been the focus of research (Thomas, 1939; Bracey, 1952, 1959; Phillips and Williams, 1984). Many people hold an idealized image of the facilities a village should offer – this image often being derived from either memories of what village life was like earlier in the century, or from media images of village life. This ideal village is likely to have a general store selling a wide variety of everyday needs; it will have a small garage and petrol station and a pub; a church that holds regular services; and also a school. Two points can be made here. The first is that this idealized view of the services available in a village is tied to a particular time in the development of rural settlements. Indeed, it is a view of village life as it might have existed in the 1950s. It usually ignores the fact the 1950s' situation is merely a snapshot of a continuously changing and, at least until recently, basically contracting scene. By the 1950s, for example, a whole series of specialist services that would have been available in many nineteenth-century settlements, such as cobblers, butchers and blacksmiths, had largely disappeared. The second is that most settlements never had the full range of services imagined. Many villages and hamlets have never had a village store, pub or school; those settlements on main roads were much more likely to have petrol stations than those tucked away down lanes. One exception to this variable distribution of services is the Anglican Church that, through its parochial system, has maintained a universal presence in the countryside, even though a number of individual church buildings have been made redundant.

In recent years there have been a number of important studies that have charted developments affecting a wide range of services in the latter part of the century. Moseley and Packham (1985) used information collected from the local branches of Women's Institutes to investigate the distribution of both fixed outlets, such as shops and surgeries, and mobile and delivery services. Major gaps in the provision of fixed services were especially apparent in the smaller settlements with less than 500 inhabitants. Over a quarter of these places lacked a daily bus, a post office or a general store. Interestingly, mobile services were also frequently lacking in these small settlements: the authors noted (*ibid.* p. 94) that 'the force of market thresholds affects mobile and delivery services just as it affects fixed services' and that the provision of mobile and delivery services increases with settlement size. This research indicates that the lack of services is especially problematical in areas where the settlement pattern is dispersed, such as parts of the Fens, the Welsh Marches and the south-west of England.

The decline of the village shop has been discussed extensively (Brown and Ward, 1990). McLaughlin (1986) notes that two main themes have dominated the rural-shopping debates – first, the decline in the number of village shops and, second, the extent to which the remaining shops are being used by the rural population. His survey results showed that for all five areas, only 17 per cent of the households lived in parishes without shops but that relatively few (39 per cent) of the households who did have a shop in their parish used it for the bulk of their shopping requirements. The survey indicated that there was a considerable degree of mutual interdependence between the poorer households and these small country shops. Moreover, the higher food prices in the rural shops had a particularly strong effect on the rural poor who spent a higher proportion of their income on food compared with the wealthier households. Clark and Woollett (1990), in their review of changes in service provision, conclude that although villages have continued to lose food shops and mobile shops during the 1980s, there seems to have been a slowing down of the rate of decline. Many shops are nowadays run by recent incomers to an area, often by people who have taken early retirement but wish to keep busy on moving to the countryside. The rate of loss of rural sub-post offices seemed to decline in the mid-1980s but recent changes in post-office policy, especially the introduction of 'community offices' that often entails a transfer to a part-time contract, indicate that closures will again become more common. Moreover, local authorities continue to search for more efficient ways of collecting their council-house rents and poll tax and this may have implications for post offices. In Dacorum District, Hertfordshire, for example, Securicor's Community Link Scheme collects poll tax and council-house rents at considerably less cost to the local authority than the post office (*The Times*, 18 August, 1990).

The availability of professional services is a key issue in rural areas. The massive changes currently underway in the organization of the National Health Service have yet to have an impact, but significant changes are already taking place as, for example, in the rationalization in the provision of cottage hospitals in small country towns. Bentham and Haynes (1984) studied the disparities in health needs and medical provision in rural Norfolk and found that those people prone to illness in the remoter rural areas were less likely to receive health care than those living in the larger settlements. A recent survey, however, paints a rather better picture in terms of the number of medical outlets. Clark and Woollett (1990) show that between 1980 and 1990 there was an increase in the number of doctors in rural areas, and more of them were dispensing medicine directly to their patients. McLaughlin (1986) found significant social-class gradients were apparent when looking at the use of health services. Fewer rural people in the lower socioeconomic groups tended to make use of doctors, dentists and opticians and those that did used them less frequently than wealthier residents, despite the fact that there was no difference in the medical needs between the socioeconomic groups.

Various studies have examined the nature of educational disadvantage in rural schools. Watkins (1979) suggested that there are pockets of educational disadvantage because of poor buildings, a low level of resourcing and low aspirations. It is well known that many of the small village schools have have been closed over the past twenty years to increase efficiency. This policy has been frequently questioned and, although there is little chance of many small rural primary schools reopening, the increasing number of 4–10-year-olds seems to preclude many further closures. The effects of the repopulation of the countryside are only now beginning to work their way through to the educational system. The number of school closures reached a peak of 127 in 1983, and by 1988 the number of closures was the lowest since 1979 (Clark and Woollett, 1990). One area where new research is required is the extent to which the increasing population of many rural areas should be taken account of in the provision of local educational facilities.

Many studies of the provision of services are based on the premise that rural areas are less well provided with services than urban areas. They are generally correct. However, recent studies of the organization of the clerical and legal professions in rural areas indicate that in one respect this premise is not true. A study of the parochial organization of five English dioceses, for example, found that although the Church has long had a policy of reducing the imbalance in manpower between the rural and urban dioceses, compared to most urban parishes, rural parishes are still proportionately over-provided with clergy. Analysis of data provided by the incumbents of 520 benefices, both urban and rural, showed that on average the most rural benefices had less than 2,000 people per benefice while in the most urban benefices there were over 9,000 people per benefice (Davies *et al.*, 1990b). It is enlightening to compare the provision of clergymen in rural areas with that of other professions. A recent study of legal services in rural Britain found that, contrary to expectations, rural areas were not badly provided with solicitors in terms of number. Indeed, it was found that the less urban a district was, the greater the chance that it would have a relatively low ratio of population to solicitors (Watkins, Blacksell and Economides, 1988). There is also a parallel here with the provision of doctors in rural areas. Phillips and Williams (1984, pp. 191–2) have noted that 'rural locations often appear to be quite adequately staffed purely in terms of the ratio of patients to health care professionals' and go on to point out that relatively rural areas such as East Anglia tend to have the smallest proportion of general practitioners having list sizes of more than 2,000 persons.

The reasons for this apparent over-provision of professional services in rural areas are complex. The location of professionals, like that of most services, is a compromise between the need to be located centrally (in order to provide an efficient service) and the need to be dispersed (in order to be close to those who require the services provided). In the case of solicitors, many factors encourage their concentration in the larger commercial, financial and administrative centres. However, a group of factors has been identified that tends to encourage their dispersal in rural areas. Some solicitors, for example, may prefer to forgo the high status associated with the larger city practices in order to gain high status within a small town community. Others may prefer to work in rural areas because of the attractiveness of the countryside and the lack of a need for commuting. The situation is rather more complex in the case of clergymen. Although some may well be affected by these sorts of considerations survey results indicate that clergymen are less able to make a free choice as to where they move as their choice is often circumscribed by the availability of benefices and the views of their bishops (Davies *et al.*, 1990c).

The ratio of providers to members of the public is, of course, only one way of looking at professional services in rural areas. The study of rural lawyers mentioned above, for example, found that there were many subtle differences in the nature of services provided by country solicitors compared to their urban colleagues. Small rural firms were less likely to provide legal services with a large social-welfare element, such as welfare and housing law, than urban firms. Moreover, the study found that people living in rural areas have to contend with a number of special difficulties that inhibit the ease with which they are able to gain access to legal advice. Access to legal services was not just limited by the physical distance from better-provided towns and cities. While there was no evidence that those in urgent need of legal advice systematically failed to obtain it, there was evidence of deprivation especially among those groups that do not have to transact routine legal business over property (Blacksell, Economides and Watkins, 1991).

Transport is, of course, the key to access to services not found in the village or parish. Rural bus services have been in decline since at least the 1960s. Following the Transport Act, 1985, the regulations covering bus services were substantially modified allowing operators to run virtually any service they wish, subject to quality licensing. The effects of this change on bus services in Clwyd and Powys are discussed by Philip Bell and Paul Cloke in Chapter 9. Whatever the effects of these recent legislative changes, however, it is clear that for the foreseeable future most people in rural areas will depend on the motor car for their transport. Rural areas in general have high levels of car ownership. McLaughlin (1986) found that as many as 71 per cent of his rural households had at least one car compared with a national figure (1981) of 61 per cent. In general, the people without cars tended to be the elderly and the poor. Half of all households with problems of access to medical services had no car and 60 per cent of the poor households had difficulty getting to doctors' surgeries by public transport. This could also be costly, especially if time had to be taken off work to make such visits. Similar very high levels of car ownership were discovered in the survey of attitudes to legal services in three remote villages in Cornwall and Devon (Blacksell, Economides and Watkins, 1991). It should not be supposed that the high levels of car ownership are simply a reflection of the wealth of rural inhabitants. People with very low wages need a car to get to work as much as the wealthy but, for the poor, the cost of running a car will form a relatively high proportion of the household budget. The wider implications of the lack of access to cars by the elderly are considered by Robert Gant and José Smith in Chapter 8.

Research implications

This chapter has emphasized the dramatic social changes that have been taking place in the British countryside in recent years, and introduced the principal themes that are explored in greater detail in the following chapters. In conclusion we wish to draw the attention of readers to the need for further detailed research into rural social change. There is clearly a need to understand more fully the range of variation in local changes in population, economy and society and how these affect the quality of life for rural residents. Of course, in many ways these changes are similar in their underlying nature to those affecting the nation as a whole, but they impact on a very different geographical environment from urban areas, with low densities of population and high variability. This gives rise to a distinctive set of policy issues, which are all the more challenging because of the rapidity of change against the background of continuity of geographical context. Clearly, this area has already stimulated a great deal of research, but there is the need for much more work.

Recent developments, such as the privatization of telecommunications, transport and water, have a number of specific implications for those living in rural areas who, for many years, have benefited from cross-subsidization of service charges. This situation could easily be defended when rural areas were seen as problem areas suffering from rural depopulation but it is now possibly more vulnerable to attack as the countryside is increasingly seen to be one of the bastions of the middle classes. A wide range of other issues demand continued investigation, including the extent of rural deprivation, the effects of changes in local taxation (including the introduction of the uniform business rate on small rural concerns), the development of new means to encourage cheap housing in the countryside and the effects of improved forms of telecommunications (such as fax) on the nature of rural employment.

One of the problems that dogs rural geographers and others working in rural studies is the lack of good data available at levels below that of the civil parish and, as shown in Chapter 12, this is one of the main reasons for the need for the social surveys that have produced so much work of high quality over the years. However, even when good-quality Census data are available, their full capabilities are frequently not used. Many of the statistics from the 1981 Census concerning small rural settlements, for example, have been relatively little used. There are remarkably few studies detailing local variations in the nature and characteristics of the rural population. Moreover, it is only very recently that some of the more detailed analyses of rural employment, occupational status and social class have been published (Errington, 1990).

The publication of the results of the 1991 Census in 1992-3 will provide the opportunity to examine the nature of the population after at least twenty years of counterurbanization. Much more information will be available at the local level compared to the 1981 Census (OPCS, 1989, 1990; Clark and Thomas, 1990). Moreover, information from wards (Local Statistics) and enumeration districts (Small-Area Statistics) will be available in machine-readable form. A further advantage is that it will be possible to obtain data for self-designed pseudo-enumeration districts, which will use post-code areas as basic building blocks. No information will be available for areas smaller than a current enumeration district, but it will be possible to modify enumeration-district boundaries in order to assist the study of local variations in rural population change. These data should prove to be extremely useful in examining the effects of long-term counterurbanization on the characteristics of the rural population.

Table 2.1   Characteristics of the inmigrants

	n	%
Restricted residents	107	22
Local-authority residents	40	8
Private tenants	67	14
Mobile residents	388	78
Young residents	265	54
Young families	140	28
Spiralists	74	15
Semi-relocators	51	10
Elderly residents	80	16
Retired	57	12
Wealthy late age	23	5

Note In addition, 43 (7%) mobile residents who had relocated over 15 years previously were interviewed.

household were identified as being significant: those in *early* and *late* age. The former comprised mainly young families, with the head of household typically under 45 years of age. However, two further sub-groups were also identified – the *spiralists* (after Watson, 1964) those who moved round the country at frequent intervals due to job-related relocations and the *semi-relocators*, who moved often short distances, usually for economic reasons, and yet still maintained full social and employment lives in their previous place of residence.

The main late age-group were the *retired*, typically over 60 years of age, who had moved to the settlement on or around retirement. In addition there was a small group of *wealthly late-age* inmigrants. Typically wealthy professionals or business executives, these people moved to the settlement not only on or around retirement but also during their 50s and early 60s prior to the cessation of full-time employment.

Residential mobility

Before considering the decision-making process of these inmigrant groups, an examination of previous research in this area will serve to identify key components of the process. The most detailed examination of household decision–making leading to residential mobility has emerged from work in urban areas. Much of this analysis has been influenced by the 1950s' work of Rossi – an advocate of the significance of *life-cycle* change. Rossi (1980, p. 6) proposed 'the major function of mobility to be the process by which families adjust their housing to the housing needs that are generated by the shifts in family composition that accompany life-cycle changes'. He argued that mobility was greatest while the family is experiencing greatest growth – young families being the most likely to move – and arose as the new family form perceived its social and spatial environment to be lacking. While Clark and Onaka (1983) subsequently questioned the confusion between *changes* in the life-cycle and *stages* of the life-cycle (changes generating mobility by altering specific household needs, stages affecting the type and frequency of these changes and of household dissatisfaction), the relationship between life-cycle and residential mobility has continually been emphasized. Thus those in the early part of the life-cycle, unrestricted by social and economic ties, are prepared to take advantage of employment opportunities requiring mobility (Lewis, 1982), while those forming their own households and families make most adjustments during the early stages (Michelson, 1977). Similarly at the end of the life-cycle, elderly people reveal a high index of mobility, as formal employment ties are relinquished and priority can be given to environmental and leisure concerns (Grundy, 1987). In addition, increasing age brings new requirements that may only be satisfied by moving nearer to sources of assistance (Harper, 1987c), while increasing evidence emphasizes the influence of income decline and the associated attraction of profit maximization through a downward move in the housing market (Murphy, 1979).

The use of the life-cycle indicator alone as a predictor of mobility is clearly inadequate. Dunnar (1979), for example, pointed out that life-cycle is closely associated with *income*, arguing that it was this latter variable that controlled mobility. *Class* has also been highlighted as an important variable (Long, 1988; Fielding, 1989), though it is still unclear whether class is the controlling variable irrespective of life-cycle stage (Leslie and Richardson, 1961; Speare, 1970). The *housing market* has also been linked to the life-cycle, with job-related moves being highest during early age (Fernadez and Dillman, 1979), rapidly decreasing as the employee reaches retirement (Swanson, Luloff and Worland, 1979).

Variables apparently unrelated to the life-cycle have also been found to be significant. Brown and Moore (1970), for example, highlighted *environmental factors*, suggesting that the encroachment of industrial, residential and commercial blight, alteration in racial or ethnic composition of the neighbourhood or in transport provision, might lead to the relocation of the household. Correspondingly, analysis of inmigration into rural areas has also highlighted environmental variables as being important location factors (Jones *et al.*, 1986). Similarly, Bell (1958) suggested that *social prestige* was a key variable in the decision process. This has more recently been substantiated within rural research by Lewis (1982), whose research in mid-Wales indicated that social motives were the prime *rationale* behind inmigration to the area.

Considerable insight has also been gained from the concept of *place utility* (Wolpert, 1966). Brown and Moore's (1970) sequential decision-making model distinguishes between the decision to move (phase 1) and the decision where to move to (phase 2), while Roseman (1971) further differentiates between the selection of the area and of the site of the new residence. The model was further refined in 1976 by Popp, who proposed that the decision to move and the selection of a new location may be independent of one another. More recently there has been an acknowledgement that individual relocation decisions are made in a *decision-making environment* (Cadwallader, 1986) and that, in order to understand the household relocation process, we must also examine the external constraints that surround it.

The decision-making process

It is thus evident that our knowledge of the household relocation process is far from complete. Much of the research has been carried out in urban environments, to the relative exclusion of the rural. Many of these studies fail to distinguish between reasons for household migration and reasons for the selection of a particular new location. While the behaviour school has been more analytical in its approach, it can be criticized for presenting the decision-making process as a logically conceptualized operation, with the household fully aware of the opportunities and constraints before them. Similarly, the relationship of the actual variables within the decision process has still to be resolved – in particular, the interaction between the life-cycle and other factors. There has also been little work on integrating macro- and micro-approaches in order to understand the process as a complex whole.

A model of the decision-making process

The study described in this chapter examines household relocation through an understanding of the migration context as a whole. Respondents were encouraged to discuss at length the relocation process rather than asked to identify key elements in the decision-making. While it is less easy to categorize 'reasons' using this technique, a far clearer picture of the process can be attained.

The discussions suggested that, while the rationale behind the arrival at the new residence might be a logically conceived operation: 'I wanted to better myself. We started off in Wheaton Aston then a similar house came up in Penkridge, and when we saw these were being built we thought it'd be an improvement. Aesthetically it's an improvement on Wheaton Aston. Just a superior village. Snob value I suppose', it was just as likely to be a jumble of unrelated ideas: 'The job brought us to the area. We liked the village. It had a good school. We didn't really rationalize things in that order – any order. There were other villages we liked but the house we eventually bought was here', or a simple feeling for place:

> We must have looked at a hundred houses . . . two or three times we nearly bought elsewhere, but it was never quite right. This house wasn't even on our list. We were here looking and I remembered I'd seen a house advertised here so we just thought we'd drive past. We had the children with us and we just stopped the car and shouted 'that's it'. We knew it was for us. Everyone said they felt at home as soon as they saw it.

or reasons that are barely admitted to the household (let alone the researcher!):

> For geographical reasons. My husband works in Walsall and both our families are in north Staffs, so here we're half-way to parents and commuting to work . . . what

actually happened was that we moved to a housing estate in town and I hated it – loathed it – so we looked for a house like this on the market, and here we are.

It is thus clear that each decision process is multi-faceted, comprising a complex pattern of reasons. As Figure 2.1 suggests, however, three broad components of the process can be determined, which can be used to understand the rationales behind household relocation:

1. The *catalyst* – that which prompts the household to change its current dwelling. This may coincide with a specific stage in the life span, arise through a personal crisis or a factor external to the household. The catalyst may directly result in the household's relocation, or may operate indirectly

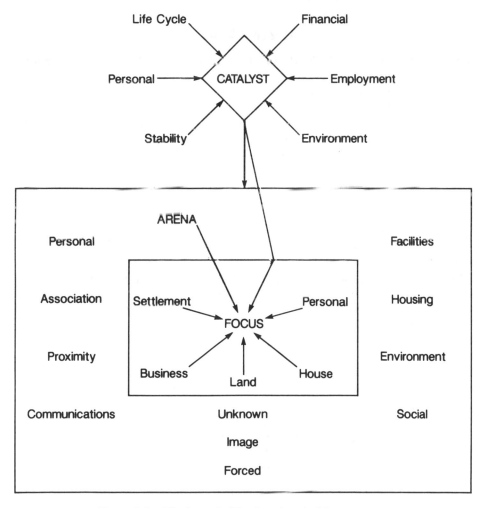

Figure 2.1 The household relocation decision process

Table 2.3   Distance of move to settlement from past place of residence

	n	Miles			
		0–10 %	11–12 %	21–50 %	Over 50 %
Restricted residents	107	49	17	10	24
Local-authority residents	40	45	25	10	20
Private tenants	67	50	12	10	17
Mobile residents	388	26	29	12	31
Young residents					
Young families	140	21	34	14	31
Spiralists	74	19	18	5	51
Semi-relocators	51	47	42	6	4
Elderly residents					
Retired	57	19	33	18	26
Wealthy late age	23	26	13	4	57
Past incomers	43	33	23	19	19

Note In addition, 12 mobile incomers had relocated the household from abroad.

Table 2.4   Number of moves made by houshold since its formation

	n	Number of moves				
		1 %	2 %	3 %	4 %	5+ %
Restricted residents	107	66	9	18	4	3
Local-authority residents	40	82	10	8	—	—
Private tenants	67	57	9	24	6	4
Mobile residents	388	31	20	13	6	30
Young residents						
Young families	140	50	26	11	4	9
Spiralists	74	5	14	19	12	50
Semi-relocators	51	58	30	12	—	—
Elderly residents						
Retired	57	28	21	19	9	23
Wealthy late age	23	4	17	13	9	57
Past incomers	43	53	13	21	11	2

restricted residents, the move to the current settlement was the first since setting up the household. This is twice the number of the mobile residents, a third of whom were on their fifth or more move. This factor appears less influenced by age – over half the retired had made only two moves since the original formation of the household – than by socioeconomic group. The higher socioeconomic groups in the case settlements include a large percentage of those whose household relocations are job related: a factor of many high socioeconomic occupations. Over half the spiralists and wealthy late age were on their fifth or more move.

Urban–rural background

As is evident from Table 2.5, the mobile and restricted residents also clearly differ in terms of the urban–rural setting of their last place of residence. While three quarters of the mobile residents have moved from an urban setting, over half of the restricted residents had previously lived in a rural setting. This latter figure is confused, however, by the large number of *tenants*, the majority of whom are tied to a rural setting by virtue of their occupation. When *local-authority* residents are considered alone they are biased towards the urban as their last place of residence. Of more interest, however, is the contrast between past and present relocators. A third of the former (relocating their households in the early and mid-1960s) had moved from a rural setting, compared with only 14 per cent some twenty years later.

Discussion

There is a complex combination of social and economic factors operating here, and clearly many of these differences will affect the reasons for the move. The higher socioeconomic groups have, by virtue of many of their occupations, a high percentage of job-related moves. These may be frequent occurrences throughout the work span and, due to the labour-market structure, may well incur relocation over long distances. The lower socioeconomic groups tend towards non-job-related moves. Their relocation is more clearly affected by life

Table 2.5   Urban – rural background of inmigrants

	n	Urban %	Rural %
Restricted residents	107	45	55
Local-authority residents	40	70	30
Private tenants	67	30	70
Mobile residents	388	81	19
Young residents			
Young families	140	86	14
Spiralists	74	85	15
Semi-relocators	51	84	16
Elderly residents			
Retired	57	79	21
Wealthy late age	23	65	35
Past incomers	43	70	30

changes, and they tend to move to areas within the vicinity known to them. An interesting contrast is provided by the *elderly* households. The wealthy late age, after a life time of moving (average number of household relocations is four), will make their final relocation to an area some distance from their previous place of residence. They are not only used to relocating such distances but their mobile life has also resulted in less attachment to one specific place. The retired in general, however, are more likely to have had few moves over their life span and, in the case areas at least, to select a retirement location in the area of their past home. We can further understand these patterns through an analysis of the decision-making process itself.

Examples of the decision process

The main relocation reasons expressed by the different groups are presented in Table 2.6. It is, however, important to emphasize that there is typically no one dominant 'reason' – rather a combination of factors that can only be understood in relation to the wider migration context of the household.

Mobile residents
Young residents

Two main catalysts had led to the household relocation for the younger residents: *employment* and *life changes*. Over half of this group had moved to the area for reasons associated with employment – typically, but not always, male employment. This arose either through the employer relocating to the region or

Table 2.6   Breakdown of main relocation reasons given by inmigrants

Catalyst	
Young residents	Employment change
	Growth of family unit
Elderly residents	Retirement
Local-authority residents	Change in family circumstances
	Employment change
Private tenants	Employment change
Arena	
Young residents	Environment
	Social
Elderly residents	Environment
Local-authority residents	Available accommodation
Private tenants	Employment
Focus	
Young residents	Settlement
	Property
Elderly residents	Type of accommodation
Local-authority residents	Available accommodation
Private tenants	Employment

the individual employee being transferred within the company. In a few cases the main employee of the household had changed employers. This reason clearly dominated the sub-group defined as spiralists, all of whom had moved to the area due to employment relocation. The second main catalyst, life change, was most commonly associated with growth of the family unit. This factor was particularly prominent among the lower socioeconomic groups, especially the semi-relocators. Their decision process put clear emphasis on the quest for cheap housing and a spacious environment. Indeed, the sudden availability of cheap rural housing (a phenomenon of south Staffordshire in the late 1970s and early 1980s) in itself provided a stimulus for relocation.

A third category of catalysts stated by the younger residents was the simple desire to leave their current (usually urban) environment, with no associated life-cycle or employment component. The most common explanations produced were environmental, social and – in the West Midlands – racial. In a few cases the household appeared to have reached a financial or social plateau, and it felt able and ready to withstand the physical upheaval of migration. These will be termed *stable moves* as they arise when the household has reached a level of equilibrium. Factors involved include the easing of financial commitments, times when children's education will not be disturbed by the move and a lack of commitments to ageing parents.

While several households mentioned *communications* and *personal* factors, the two main factors determining the search arena for the younger residents were *environment* and *social*. This process was underlined with two broad preconceptions about a rural setting – 'peace and quiet' and 'a sense of community': 'We came to this area because of my husband's job. We came to the village hoping to find village life – which we haven't – it's been very disappointing, there's no village life at all, in fact quite the opposite'. Several of the lower socioeconomic households, however, displayed a different rationale, in particular those identified as semi-relocators. For these the search for 'community' or 'village life' appeared singularly lacking as a motive for inmigration; indeed, few of this group appeared to have a clear conception of life in a small settlement, primarily perceiving the rural zone as possessing relative advantages of housing and recreation facilities in comparison with their urban home. In contrast to this, many of the spiralists drew on past experiences in selecting the search arena. Though the majority had little or no personal knowledge of the area, their continuous relocation, by which they had encountered varied milieu, had given many of the households valuable insights, and they were thus able to make sophisticated judgements on the basis of their knowledge of the process.

The focus of the search was polarized between those selecting a particular *settlement* (46 per cent) and those choosing a specific *house* (43 per cent). The high proportion of these residents deciding to live in modern estate housing (50 per cent of the study's sample) demonstrates that priority can be given to a settlement containing appropriate property, and then the decision taken between several suitable houses: 'It's the nicest village in the area . . . we like the joys of village life . . . we decided we wanted to live here and the house came second'. It is worth noting that this generally urban method of choosing a settlement and then selecting property within its confines is now able to operate in rural zones. This type of relocation was particularly prominent among the semi-relocators, who were often attracted to a settlement purely because it contained a large number of types of accommodation perceived as appropriate. It was also a strategy favoured by many spiralists.

Past incomers

Before turning to consider the decision process of the older residents it is worth discussing the relocation of those residents who moved to the settlements in the first wave of immigration some 20–25 years ago.

The catalyst behind their relocation appeared very similar to that revealed by the other young incomers – primarily life changes and employment. However, both the choice of search arena and the focus reveal a far stronger emphasis on *personal connections* and *personal associations* than did the other incomers. Indeed, a quarter of the past incomers recorded a personal motive as of prime importance in the decision process, compared with only 12 per cent of the more recent incomers.

Thus, while the relocation catalyst seems to have altered little over the years, the obtaining of property has undergone a substantial change. As the demand for housing has increased, so the rather *ad hoc* individual method of aquiring land and housing through the use of personal aquaintances has been replaced by a more impersonal, regulated procedure. This is, of course, spatially reflected in the shift from small infilling and isolated building, scattered around and within the settlement, to mass-produced housing estates in preordained locations.

It is possible, however, that the fact that these inmigrants have remained in the settlement for over 15 years is related to their original reason for inmigration. We may thus be examining a particular type of inmigrant with a specific relocation decision-making processs.

Elderly residents

As was described earlier, the majority of the elderly mobile residents had moved to the settlement on or around retirement. A quarter of these elderly residents, however, the wealthy late age, revealed strong similarities in wealth, socioeconomic group, lifestyle and house type. Their relocation to the settlement differed from the rest of the late-age residents, both in terms of decision and process.

The main catalyst for the retired resident was *retirement per se*, with two thirds perceiving the concept 'retirement' as their prime motive in the decision process. Thus economic, social and psychological tensions that accompany this physical and symbolic *rite de passage* are released in the spatial upheaval of the household through migration. *Economic* tensions were often the result of direct financial pressure: 'We came here for financial reasons, we had a big house and garden and needed somewhere smaller'. *Psychologically*, there was often a simple desire for a change. Many perceived retirement as representing physical movement and the alteration of the life state. For these elderly residents, 'we retired' was the complete and only explanation required to account for their migration.

This is in contrast to the wealthy late-age residents, many of whom were able to stagger retirement and prepare both physically and psychologically for the eventual abandonment of full-time employment. The key period for this group was *approaching retirement*. As employment schedules and commitments altered with late career development, so some professionals and executives and those in the services took up the opportunity to select a final rural home within commuting (often weekly) distance of their employment, and there establish the household prior to full-time retirement. Alternatively, the rural property may be purchased as a weekend home several years before permanent residence is undertaken.

Environmental considerations appeared important across the spectum of elderly residents in the selection of a search arena. One quarter of the main retired group, and over half of the wealthy elderly residents, selected this factor as the *main* determining factor. Also of interest is the fact that the former associated environmental factors with a 'sense of community', while for the wealthier residents the association was typically with a 'particular lifestyle'.

There was a marked difference also in the focus component of the decision process. Over two thirds of the wealthy elderly residents had moved to their current settlement due to the discovery of a *particular property*. The houses occupied by these residents are large: a typical example may have five or six bedrooms, two bathrooms, four reception rooms, kitchen and utility facilities, with adjacent servants' quarters, now usually part of the family living quarters, although resident house-keepers and gardeners were in evidence, and several acres of grounds. Former manor houses, vicarages, farmhouses and converted cottages are among the property considered appropriate.

Ownership of such property was a status symbol, an overt recognizable sign of wealth, and local friendships were selected from the owners of similar accommodation. These houses were thus purchased for reasons other than practical use, for rarely did the inhabitants occupy all the space available – their families having departed, it was common to find the couple living alone.

The main retired group, however, typically followed the procedure of the younger residents, and searched for a *type of accommodation* rather than a specific property. The village might thus be selected for personal, social or environmental reasons, and suitable housing chosen from that available within the settlement.

Restricted residents

As was indicated in Table 2.1, the restricted inmigrants revealed a very different decision-making process from that of their mobile counterparts. This was so linked to the mechanisms associated with the specific type of tenure of the property they inhabited that other socio-cultural factors had relatively little impact on their decisions. For these, employment provided both the catalyst and focus component of the decision process. This was primarily agricultural, though some were occupied in domestic and maintenance work. Although just under 25 per cent had moved to the case settlement from over 50 miles distant (finding the positions through formal advertising procedures), the majority of those in tied accommodation were born and raised in the general vicinity.

Private tenants

Of those in private rented accommodation over three quarters lived in property tied to their current or former employment. For these, employment, which was primarily agricultural, though some were occupied in domestic and maintenance work, provided both the catalyst and focus component of the decision process. The majority of those in tied accommodation were familiar with the locality, and often associated in some way with the settlement and their employers, finding their current job through personal connections, rather than through the formal mechanisms of the labour market.

Local-authority tenants

The second group of restricted incomers faced two main procedures of moving within the local-authority sector: they could apply to the local authority for a transfer or privately arrange a mutual exchange with another local-authority resident. All those interviewed had received formal transfers through their local authority. The group fell into two broad categories: those arriving from outside the region, mainly for employment, and those moving within the region, with deteriorating conditions, health and a deficiency or surplus of accommodation being given priority for transfer in both study areas.

   The catalysts for this group were similar to those of the mobile incomers. Life changes, in particular change of family circumstance with *marriage* and *change in household size* were the most common, followed by retirement and employment. However, a not insubstantial number also described how *ill health* and the *deterioration* of property had also led to their being rehoused:

> I was so poorly. I'd been poorly for a while and they'd do nothing about it. Doctor said it was the damp. Oh, it was so damp . . . but they wouldn't come round . . . and then one winter my daughter came round and found me lying on the floor and she got the social worker, and they moved mè out, and I came here.

   The arena is controlled by the distribution of local-authority property and its allocation; the focus dependent on available accommodation within the arena.

Conclusion

There is a clear correlation between those factors identified at both the macro- and household levels. One of the most significant to emerge is the regional contrast between the two counties. The households in south Hampshire gave a far higher proportion of employment reasons, supporting the findings at the macro-level that the main engine of growth in this area is employment based: counterurbanization is thus primarily an employment-led phenomena in this county. By way of contrast, in south Staffordshire many more households gave the attraction of housing in the countryside as their primary rational for relocation, in many cases the householder retaining his or her employment in the conurbation. As suggested by the Staffordshire survey, counterurbanization appears to be primarily housing led in this part of the county.

There are also indications that urban–rural migration may have undergone a fundamental change in its character over the past two decades. Though the evidence from this survey is based on a small sample, and given the qualifications mentioned earlier, the decisions made in the late 1950s and early 1960s were typically based on personal and prior association, and very individual in character. They have been replaced by mass movements attracted to large developments, which are typically unknown to the household prior to the relocation, the move being based on abstract preconceptions of the area. This change has been further emphasized by the increasing possibility to undertake urban-style moves in the countryside. The indications from the study are that settlement selection is increasingly replacing property-based selection in these areas for many of the inmigrants.

At the household level, class differences account for the greatest variation in household relocation. Higher socioeconomic groups were more likely to relocate over long distances than lower groups. As employment and educational aspirations were less likely to be satisfied locally for the former group, they were accustomed from an early age to moving long distances to adapt to life changes. The lower socio-groups (with their greater tendency to find education and employment within the area) were also predisposed to select local marriage partners, thus further cementing their attachment to the locality. Thus, while age accounted for individual variation in the decision process, at all ages life-stage changes resulted in longer distance moves for the higher socioeconomic groups. Correlated with this were the higher percentage of employment-based moves among the higher socioeconomic groups.

Analysis of the household relocation process has indicated that counterurbanization comprises a complex system of individual decisions. Most importantly, the survey has suggested that it is not one mass movement but is rather locality based, and will vary in its process between different areas. Beyond this are indications that counterurbanization emerged as a result of pressure on urban land, encouraging the development of the area beyond the city. These developments were attractive to the population because they arose, first, at a time of growing environmental awareness, which replaced the image of the rural as backward with a positive social image; and, second, with growing mobility among all socioeconomic groups both in terms of travel and in access to private property. Many of those involved in the counterurbanization movement are the first generation of their families both to leave their kin-based locality and to become owner-occupiers. Many studies have generalized the counterurbanization movement, extrapolating their findings from macro-level data alone. This study has shown that it in order to understand the movement, it is essential that the process is considered at both micro- and macro-levels.

3

MIDDLE-CLASS HOUSING CHOICE: CHANNELS OF ENTRY INTO GOWER, SOUTH WALES

Paul Cloke, Martin Phillips and David Rankin

Middle-class fractions and the 'colonizing' of rural areas

Researchers have long taken an interest in the planned development of new housing in villages and small towns. Much of this interest has focused on the interrelationships between planning and housing development, dealing, for example, with issues such as central government policy change (Forrest and Murie, 1988), the rural settlement policies of local authorities (Cloke, 1979; 1983a) and contemporary mechanisms aimed at providing low-cost housing for local people (Shucksmith, 1981; Phillips and Williams, 1982; NAC Rural Trust, 1987). In each of these three examples, the tendency has been to examine local housing issues in order to highlight the plight of low- or no-income local social groups whose access to local housing markets has been increasingly restricted because of rising prices, external competition from adventitious inmigrants and the constraints in housing supply due to strict planning policies for settlement conservation. This focus on the need for specific mechanisms to provide affordable housing for local people is (as shown in Chapter 4) entirely appropriate and important, but has overshadowed somewhat some other significant issues concerning the link between the supply of housing and social recomposition in rural communities.

Chapter 2 explored the reasons why people move to the countryside. This chapter seeks to throw a little light on the links between housing provision and the colonization of particular rural localities by specific groups of people or class 'fractions'.[1] Analysis of this topic has been characterized by a number of truisms: restrictive planning policies *cause* an upward spiral in house prices; occupancy of houses thereby becomes restricted to those social groups who can afford the new prices; the village concerned becomes 'gentrified' (this may or may not involve an upgrading of the housing stock via extension or refurbishment); the new rural middle class are immediately engaged in social conflict with the 'working-class locals'; and local political power gradually accrues to newcomers. Although this sequence of events seems to bear some semblance of truth for some rural places during some time periods, it is far too simple a backcloth against which to set contemporary studies of changing class and culture in rural Britain. What is required is both a much more sensitive appreciation of the specific impacts that different middle-class fractions have in the social, economic and political dealings of rural life, and a realization that *intra*-class conflicts have become important in contemporary social change in rural areas. Some of these themes have been reviewed by Cloke and Thrift (1987) in terms of the social relations that have given the members of particular service class fractions[2] specific powers with which they can make themselves different from other class fractions. Three such social relations were highlighted:

1. The increasing complexity of the division and organization of labour, particularly in large corporate and State bureaucracies, in which certain labour fractions can establish stable and protected careers.
2. The capture of key skills of organization and bureaucracy by fractions of labour power with particular educational credentials.
3. The appropriation of the means of consumption by these same fractions who are thus able to create distinctive cultural lifestyles.

It is this third social relation that is of importance to our understanding of rural housing as a channel through which the colonizing power of particular class fractions is exercised. Thrift (1987) has shown that service classes have manifest residential preferences towards rural areas; they have a strong cultural affinity towards the rural idyll and the set of values said to emerge from rural communities of the past; and they wish to conserve this idyll while adapting the places concerned to present-day needs of consumption. In a geographical context this has meant that certain rural localities (particularly in the south of England) have been attractive to these class fractions. They contain the right kind of housing stock and are near to important theatres of consumption, notably small historic towns, which allow the consumption-based lifestyles of the service class to flourish.

Given this specific confluence of events, these rural locales do experience a spiral of house-price inflation. Such price increases reflect the fact that the housing stock has assumed a significant *positional* component because of its character and location (Scitovsky, 1987). The retention of this positional component essentially involves a restricted supply of housing so that prices remain high. Additionally, in some cases the location of a particular village is of higher value than the available housing stock. In the positional housing market we can therefore recognize a selective partnership between the colonizing classes and housebuilders (Barlow and Savage, 1986).

The colonization of particular rural locales by members of the service class is, then, also tied up with the role of land-use planning mechanisms. The limiting of housebuilding in smaller scenic settlements to a few high-quality, design-conscious, high-price houses serves the purpose admirably. Larger-scale growth can be accommodated elsewhere. In many cases these planning policies are reinforced by an increasing infiltration of service-class representatives into local politics. Even without this formal representation, service classes have been able to appeal to local values of conservation and heritage in order to maintain their colonization by keeping other classes and class fractions out of the locality (having already priced out many others). These aspects of class conflict are further discussed in Cloke and Thrift (1990).

In the context of this chapter it is sufficient to suggest that preliminary work on service classes presents an illustration of the way in which one specific class fraction can colonize a particular rural locale. Housing – either existing or selective newbuild – represents the *channel of entry* for these classes and thereafter is important as a commodity of tremendous positional importance. There is no implication here, however, that all class fractions in all locations behave exactly in this manner. In recent discussions of the links between rural localities and social recomposition there has been a danger of establishing a new superficial orthodoxy of rural social and cultural change based loosely on the seemingly ubiquitous service classes (*ibid.*). Clearly, such a new orthodoxy is as open to the ecological fallacy as is the previous correlation of 'middle-class' and 'newcomer' groups. There is also a very real danger of under-estimating the effects of specific localities in the interrelations between the central State, the local state and housing capital in fashioning different channels of entry for different fractions of the middle class into particular rural places through the housing market.

The remainder of this chapter examines the channels of housing entry in one such locality – the Gower peninsula in south Wales – and in some respects uncovers significant differences in the configuration of interrelations described above with reference to the service class in rural southern England.

The housing–planning relationship in the Gower

The potential for colonization

The Gower peninsula is located in the county of West Glamorgan, west of the string of urban centres that have collectively become known as Swansea Bay City (Figure 3.1). As such it is part of an area that straddles the cultural divide between the Anglo-Welsh south and the rural heartland of west and mid-Wales. With the restructuring of the Swansea Bay City economy away from the larger traditional industrial employers (although some important plants remain) there have been strong attempts by local authorities and the Welsh Development Agency to attract into the area employers in the administrative, commercial, scientific and high-technology sectors. Two factors relating to land-use planning have been used as part of the marketing exercise that has been adopted. First, the West Glamorgan Structure Plan acknowledges the need to consider providing land for those employers requiring exceptional environmental conditions or prestigious locations. Second, there has been a strong emphasis on the quality of life available to in-coming entrepreneurs in such scenic areas as the Brecon Beacons National Park and the Gower Area of Outstanding Natural

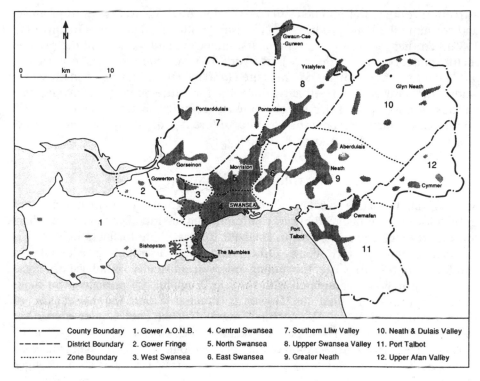

Figure 3.1   West Glamorgan Structure Plan zones

Bcauty (AONB). Thus an integral element of the economic image of the area has bccn the availability of appropriate housing in appropriate rural locations for the different middle-class fractions who might be attracted in either as entrepreneurs or as key managerial workers. Such a policy has obvious relevance to the Gower as a locale for potential colonization.

Since the 1960s the majority of the population of Gower has worked outside the area. In 1921 only 20 per cent of its working population travelled to work outside the area. By 1971 this had increased to 66 per cent, with Swansea forming the primary focus of travel to work (Swansea City Council, 1978). The pressure on the Gower as a residential locality does not, however, stem only from the existing and potential labour market of the Swansea area. In relative terms, house prices make the Swansea area less expensive that other areas in South Wales (such as Cardiff and Newport) and other regions of southern England. It can also be envisaged that residential pressure could therefore stem from a westward drift of some people connected with the Cardiff, Bristol and even London labour markets. These could be further fractions of inmigrants or, alternatively, people with strong links with Wales who are seeking a placc they know, where they can retain certain cultural characteristics of Welshness, *and* take account of a chcaper housing market. In addition, Gower's tourist role may attract entrepreneural middle-class colonization by those who wish to produce and consume the same product, i.e. 'tourism' (Williams, Shaw and Greenwood, 1989). To these groups of entrepreneurs, managers and workers in the private sector must be added those whose work is connected to the very strong *public-sector* presence in the Swansea area (and indeed in Cardiff), and those for whom the Gower is an attractive location for retirement. Collectively, the various groups of residents constitute a complex heterogeneous set of middle-class fractions, potentially quite different from those in the service classes in southern England.

The role of local planning

Some of the distinctive social relations that underlie housing and planning issues in the Gower are influenced by very different administrative circumstances than those pertaining to rural areas in England. Central State business in Wales is channelled through the Welsh Office, and some discretion is available in planning matters to vary the timing or even substance of advice to local authorities in Wales compared with those in England. Of perhaps even more direct importance is that the Gower is situated within the jurisdiction of Labour-controlled West Glamorgan County Council and Labour-controlled Swansea City Council. These political leanings are unorthodox for most 'rural' areas, and lead to the expectation of not only ideological and political turbulence between the central and local states over Gower issues but also a strongly interrelated scheme of planning between the city of Swansea and the Gower.

At least the second of these expectations is borne out by the planning strategy for the area west of Swansea employed in the Structure Plan (West Glamorgan County Council, 1980). Figure 3.2 demonstrates the three main planning arenas concerned, as well as the proposed 'green wedge' policy, which transcends the policy boundaries. In each case a series of clear and interrelated planning aims was established.

Gower AONB

Over the previous two decades there had been strong suburban pressure in the eastern area of the Gower, such that the rural population grew by 30 per cent between 1961 and 1971 and by 12 per cent between 1971 and 1981. As well as the continuation of this potential for suburban spread, there is sporadic but strong pressure for prestigious developments on infill sites in many Gower villages.

The policy of the Structure Plan, as implemented in the decisions of Swansea City Council through their Local Plan (1989) is to prevent westward urban sprawl from Swansea into the Gower, to preserve a village landscape and to attempt to ensure the survival of villages as balanced communities. The policy on village landscapes is said to be directed at conserving a highly valued 'village character' (*ibid.* p. 76). The precise nature of this village landscape is elusive to capture in definitions although, according to the Swansea Local Plan (*ibid.* p. 76), it includes 'the lie of the land . . . innumerable social influences of past centuries . . . vistas of countryside and glimpses into yards . . . with houses of

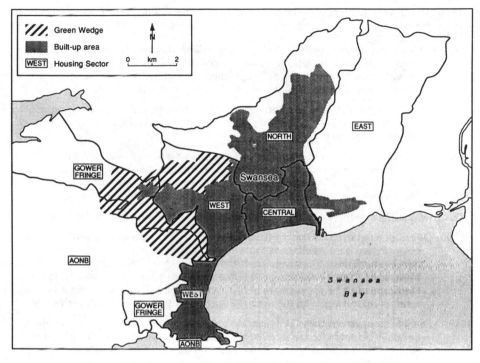

Figure 3.2   Planning arenas in the Gower and west Swansea

different styles and public houses, village shops and the church'. The image created has important potential impacts on channels of entry into the area. First, it appears to coincide with the preferences of service-class members as described earlier. Second, land-use policies pursued by the local planning agencies are helping to produce housing stock consistent with this image. For example, in the Swansea Local Plan (*ibid.* p. 77) it is declared that the main thrust of policies for villages in the Gower AONB should be 'aimed at the implications of development policies for visual form and character'. Table 3.1 provides a brief summary of policies in the AONB that are directed at this aim. As indicated, a concern with the aesthetic form of the village encompasses not only 'village policies' and landscape policies but also the agricultural policy seen as part of a strategy to ensure the continuance of balanced communities. Balance here is defined in terms of negatives – *not* merely commuter settlements, *not* dominated by second or holiday homes, etc. – rather than positive characteristics. These planning aims are being realized through the use of AONB status to ward off big housing developments in the eastern villages of the Gower. The power to refuse such applications on these grounds has by now been well tested through past appeals, and there now seems to be a tacit acceptance of this by planners and housebuilders alike. Moreover, the very proximity of these eastern villages to the city of Swansea itself allows the local authority to divert development pressure into other available sites in the city.

Table 3.1   Examples of plans relating to 'village character' within the Gower AONB

Policies description

GL2   Allow only development of the highest standards that will enhance the character and economy of the Gower and that is consistent with adopted social and environmental objectives. Such development will need to achieve the highest standards of architectural design, landscaping and siting

GL3   No development will be permitted that will be detrimental to the character of a conservation area, of a building listed as being of special architectural or historical interest

GL5   Developers required to retain existing hedgerows and trees wherever practicable and to include appropriate means of enclosure and new planting in all developments. Appropriate means of enclosure include natural banks, hedgerows and walls constructed with a 'natural, locally found' stone

GA2   Require the sympathetic siting and design of farm buildings in the landscape

GV1   Development permitted only within defined Village Study areas. Infill projects to be permitted only when in accord with provisions of the Village Studies. Such development must 'positively contribute to the character of the village', 'Its form, scale, elevational design, relationship with adjoining building and landscape must be appropriate to the location' and should be 'compatible, in terms of operation, with the over-riding residential nature of Gower villages'

GV2   Outside areas covered by GV1 and GV3–8, residential development will not normally be permitted, but 'reuse of derelict, redundant agricultural buildings and cottages of merit will be encouraged provided it would bring about an improvement in landscape or village scene'

Notes GL Gower Landscape Policies; GA Gower Agricultural Policies; GV Gower Villages Policies.

(*Source*: Swansea City Council, 1989 (draft).)

Channels of middle-class entry into this area, then, are restricted to the existing housing market, and to small amounts of prestigious infill newbuild. Larger-scale development pressure has been diverted eastwards into other Swansea housing zones, and conditions in the Gower therefore seem ripe for colonization by particular class fractions for whom housing is both a channel of entry and a positional good. It should be stressed here that the process of colonization will be variegated within the Gower. Some settlements will offer more prestigious locations and housing characteristics than others; the incidence of appropriate channels of entry may occur more at particular times in some places than others. Nevertheless, the overall planning policy of development restrictions aids the potential colonization process. Interestingly, the flipside of local planning policy – attempts to ensure 'balanced' communities – has met with strong opposition from local groups in certain Gower settlements. Plans for local village workshops are perceived by planners to have been strongly opposed by local residents as were specific proposals for the release of council-owned land in the village of Scurlage for a housing-association development of affordable housing.

Gower Fringe

This is the part of 'rural' Gower that is most vulnerable to urban development as it lies outside the AONB. The Structure Plan employs a *green-wedge* policy in that part of the fringe abutting west Swansea, proposing that strict controls on development should be implemented therein. Thus, while satisfactory proposals for infilling are permitted, bulk development pressure is again diverted to more accommodating parts of the city of Swansea.

There is very strong developer interest in the Gower Fringe, but currently the availability of prestigous development sites in west Swansea, the Maritime Quarter of Swansea and (potentially) new riverside sides following the barrage development, are to some extent satisfying housebuilder demand for prime building land. Control of development in the fringe is less well tested by past appeals than that in the AONB, however, and certainly there is less tacit acceptance by housebuilders of development controls over these sites. If central government continue to uphold increasing numbers of appeals against refusal to develop, then the vulnerability of the Gower Fringe to development may well be exposed regardless of local-authority opposition.

West Swansea

Outside the designated green wedge, all available sites for housing development in this area will be quickly used up. Indeed, most of these sites are already landbanked by developers. This has been the major area into which development pressure on the Gower and the Gower Fringe has been diverted. Once sites here become scarce, the important issue will arise as to whether Swansea City Council will be as successful in diverting pressure from the Gower into the more distant parts of Swansea (where suitable land is available) as it has been in securing the tacit acceptance that sites in West Swansea are acceptable in lieu of development in the Gower.

The other major factor in the West Swansea area is the question of whether the green-wedge policy can be implemented effectively. Scarcity of sites outside the green wedge will reinforce the status of green-wedge land as prestigious and desirable for housing development. If attempts by housebuilders to secure permission to develop inside the green wedge are successful, then a major precedent would be set that, in time, would create important difficulties in opposing development in the Gower Fringe where green-wedge policy is the basis of the council's attempt to control development.

Welsh office intervention: – circular 30/86: *Housing for Senior Management*

The Welsh Office circulars

The planning policies examined in the previous section reflect a complex set of relations between residents, potential residents, housebuilders and local authorites. The planning arena of the Gower was, however, subjected to further complexity due to an attempt by the Welsh Office to advise local authorities to make special provisions for certain class fractions to gain access to appropriate housing in rural areas. This intervention was not on behalf of low-income local people, but rather on behalf of 'senior managers' and it represents a very specific policy tool designed to achieve positive discrimination in favour of some of the middle-class colonizing fractions discussed in this chapter.

In June 1986, the Welsh Office published a circular (no. 30/86) entitled *Housing for Senior Management*, the intention of which is evident from the following extract (para. 1):

> The Secretary of State has had his attention drawn to the possible disincentive to business people who might otherwise be prepared to invest and develop in Wales, brought about by the absence of an adequate pool of houses attractive to and suitable for senior managers and senior technical staff. The difficulty of finding sites for new homes in attractive locations suitable for an individually designed residence commanding the privacy and space which many executives in business and industry expect and are often able to find in other countries needs to be addressed. The Secretary of State . . . therefore asks each local planning authority to review their policies on the control of development with the object of introducing sufficient flexibility into their strategy on developments in and adjoining towns and villages . . . to permit additional sites for such single new houses or low density groups of houses so as to ensure that there is the pool of houses and sites for houses necessary for the encouragement of new commerce and industry.

In September 1987, another circular (no. 44/87), under the same title, reported on the responses of local authorities to circular 30/86 but reaffirmed the basic principles contained therein, stating that (para. 6) 'The Secretary of State and his inspectors will, in dealing with planning appeals, have regard to the objectives described in this Circular'. Circulars 30/86 and 44/87 raise very important general questions about the use of planning mechanisms to favour particular social groups, and are particularly poignant in view of the absence *at that time* of the measures that have subsequently been taken to provide land for low-cost housing in rural areas. More specifically, it is clear that the implementation of the objectives of the circulars could have an important impact on the channels of entry for particular middle-class fractions into an area such as the Gower. As circular 30/86 stated that it was not expected that the intended flexibility should be implemented in national parks, AONBs, Heritage Coast, National Nature Reserves and SSSIs, the Gower AONB would be unaffected by these measures, but the Gower Fringe would be a prime target for developers wishing to use the circular to provide new channels of entry into the area. The Gower AONB was not immune from the provisions of the circular, however, as the attention of local planning authorities was also drawn to a 'related problem' (*ibid*. para. 5): 'the need to ensure that attractive residential areas are not spoilt by insensitive infilling'. The inference to be made here is that prestigious and exclusive villages suitable for colonization by the managers and other class fractions favoured by the circular should be preserved for that purpose. Hence when colonization takes place through access to existing housing stock, the subsequent process of entrenchment by the class fractions concerned is specifically aided by this invitation to strengthen planning controls over new development. Such measures could certainly raise the exclusively and positionality of key housing areas within prestigious parts of the Gower.

The Swansea Local Plan

The impact of the measures contained within circulars 30/86 and 44/87 can be more fully assessed in the context of the attempts by the Welsh Office to persuade local authorities to change their residential policies accordingly. One of the key test cases was the public local inquiry held in December 1988 and January 1989 into the formal objections made to the Swansea Local Plan (Swansea City Council, 1989).

The background to the public inquiry is that, before putting a local plan on deposit for inspection, the local district planning authority has to obtain from the county council a certificate stating that the local plan conforms with the Structure Plan. Swansea City Council received their certification of local plan policies (mark-one version) from West Glamorgan County Council on 27 November 1987, but subsequently the Welsh Office insisted on modifications to Structure Plan policy H4 to accommodate the measures confirmed by circular 44/87 on housing for senior management. The original policy H4 (vi) named specific Gower Fringe villages and stated that development would be restricted to that which does not adversely affect the character or amenity of the village and would normally take the form of limited infilling. The Secretary of State added the clause 'and small scale rounding off as well as minor extensions of existing settlements for low density development' in order that the policy should be adequate to allow for the provisions of circular 30/86. At the local-plan public inquiry, the MP for the Gower, Gareth Wardell, noted that 'their modification to Policy H4 (vi) of the Structure Plan against the strongly held views of the County Council side-stepped public consultations, meaningful explanation and debate. They did not appear at the Examinations in Public and thereby denied other participants an opportunity to cross-examine' (para. 13.14). The fact that this is a Labour MP supporting a Labour county council against the actions of a Conservative government Welsh Office perhaps explains the language of these complaints, but does not dilute the strength of the procedural accusations levelled at the Welsh Office over this issue.

Following the modification of Structure Plan policies, the Secretary of State for Wales on 19 July 1988 made a section-14 direction on Swansea City Council requiring them to modify policies for Gower Villages in line with the new Structure Plan policies, embodying the circular 44/87 measures. The covering letter suggested that the local plan could not be formally adopted until such modifications were implemented. Discussions followed between the city councils and the Welsh Office and, in due course, modifications were drafted to the Gower Village policies (the mark-2 version), which were placed on formal deposit along with the rest of the draft local plan between 3 October and 14 November 1988 (see Table 3.2). At a meeting of the city council on 23 November 1988, a mark-3 version of the Gower Village policies was agreed,

People in the Countryside

Table 3.2   An illustration of the draft local-plan policies for Gower Villages (GV3–8)

Policy GV7: the mark-2 version
It is the policy of the City council generally to resist extensions of Dunvant into the open
countryside and in particular to avoid coalescence with Killay and intrusion into the Llyne
Valley. Although the possibility is not precluded of constructing a limited number of
single dwellings of quality strictly in accordance with the requirements of Welsh Office
Circular 30/86 at appropriate locations adjoining the settlements, and subject to overall
safeguards relating to protection of character, amenity and high quality agricultural land.
   In assessing such proposals particular attention will be paid to the prominence and
visual impact of a site, its relationships to the spatial structure of the village and the extent
to which it can be screened and contained, and to the design suitability of the proposed
building.
   Within the defined area appropriate infill schemes will be permitted particularly where
they promote improvement and where they do not prejudice local landscape features. An
environmental enhancement scheme in the area of the old Dunvant station will be
implemented.

(*Source*: Swansea City Council, 1989 (draft).)

which took account of public objections and which was to be used for the
purposes of the forthcoming public inquiry. In the example of policy GV7
detailed in Table 3.1, the words 'STRICTLY IN ACCORDANCE WITH THE REQUIRE-
MENTS OF WELSH OFFICE CIRCULAR 30/86 AT APPROPRIATE LOCATIONS ADJOINING
THE SETTLEMENTS' were replaced by 'NOT IN GROUPS AND OF INDIVIDUAL AND
EXCEPTIONAL STANDARD OF DESIGN *MEETING A NEED WHICH CANNOT BE PROVIDED
FOR BY THE EXISTING HOUSING STOCK*, BE AT APPROPRIATE LOCATIONS WITHIN AND
ADJOINING THE SETTLEMENTS'. The Welsh Office were advised of these changes
and responded by letter dated 12 December 1988 that the phrase 'meeting a
need which cannot be provided for by the existing housing stock' did not meet
the terms of the section-14 direction and should be deleted from each of the
Gower Village policies.

The public inquiry

Formal objections to the draft policies were heard at the public inquiry in
December 1988 and January 1989. Most of the objections focused on the policies
for villages in the Gower Fringe, and the inquiry became a forum for analysing
the effects of amending these policies in line with circular 30/86, which in turn
became the (unofficial) nub of the inquiry. The 125 objectors were represented
by seven sets of aural evidence at the hearing (Table 3.3). A review of some of
the points raised by Gareth Wardell, MP, serves to summarize the scope of
objections raised (Table 3.4).
   Thus the general fears that these policy modifications would open up Gower
Fringe villages to substantial development pressure with consequent negative
effects on conservation (and investment and positional goals of residents) was
accompanied by a strong and cogent attack on the circular 30/86 mechanism of

Table 3.3   Presenters of aural evidence at the public inquiry

1.   Gareth Wardell, MP (Gower)
2.   Gower Society
3.   Kittle Residents Association
4.   Pennard Community Council
5.   Council for the Protection of Rural Wales
6.   Bishopston Community Council
7.   Society of Architects in Wales

Table 3.4   Review of points raised by Gareth Wardell, MP

1.   The amended policies for Gower Fringe villages are a threat to the conservation of the rural character of the villages, and to those policies that seek to prevent the urbanization of land between Swansea and the Gower AONB
2.   Personal parliamentary questions show that only 4 out of 54 responses to the Welsh Office consultation on circular 30/86 had been favourable. These were from

  • *the Land Authority of Wales*, who already have powers that enable them to acquire land for housing special groups if there is a need;
  • the *Welsh CBI*, who were representing the interests of their various clients including the housebuilders;
  • the *Welsh Development Agency*, who subsequently were said to have conceded that the issuing of circular 30/86 was 'an error of judgement'; and
  • *Winvest*, a part of the Welsh Development Agency

3.   At a meeting at Bishopston on 5 June 1987, constituents of Mr Wardell were told by the then Secretary of State that circular 30/86 would not be relevant in the Gower. This raises the question of whether the Gower as a place stops at the AONB boundary
4.   Circular 30/86 discriminates in favour of particular social groups at the expense of the rest of the community. This is seen as unfair and improper
5.   If a policy 'loophole' is left so that circular 30/86 can be activated, housing developers will deem such sites as immediately available for building new residences for upwardly mobile existing residents of Swansea, rather than restricting sites to incoming industrialists who will bring a significant number of new jobs. This assertion was illustrated by a rash of recent planning applications in the Gower Fringe using circular 30/86 as a basis
6.   The policies reflecting circular 30/86 were contrary to other Welsh Office advice, particularly

  • circular 1/85, which states that planning is concerned with land use rather than who occupies dwellings;
  • circular 47/84, which encourages local authorities to make land available for home-ownership on as broad a basis as is possible;
  • circular 43/84, which highlights that it is inappropriate to formulate policies where planning applications will be dealt with on the basis of personal need or characteristics; and
  • circulars 40/80 and 20/80, which confirm the need to protect rural environments from urban sprawl and ribbon development

7.   Parliamentary questions show that the Department of the Environment has no intention of issuing a circular equivalent to 30/86 in England
8.   Circular 30/86 appears to contradict current planning case law. In the case of Great Portland Estates, plc *v*. Westminster City Council in 1984 (*All England Law Reports*, 744) Lord Scarman ruled that (emphasis added): (cont.)

Table 3.4   con't.

---

although the formulation of planning policies and proposals was concerned with the character of the use of land rather than the particular purpose of the particular occupier, in exceptional or special circumstances the personal circumstances of an occupier, personal hardship or the difficulties of business which were of value in the character of the community were not to be ignored in the administration of planning control, *but such cases could not be made the subject of a general policy and could duly be considered as specific objections to a general policy.*
It follows that the insistance that Gower Village policies be amended to meet the requirements of circular 30/86 is *ultra vires*

9.   The Welsh Office was seen to have consistently discriminated against West Glamorgan County Council, and thereby against Swansea City Council by applying circular 30/86 to the Structure Plans and local plans in the area when it had not enforced similar modifications elsewhere in Wales. In addition, the need for and relevance of circular 30/86 had not been justified.

---

special privilege for senior managers and equivalent class fractions. Equally, there was strong support for Swansea City Council's mark-3 policy clause restricting circular 30/86 housing to situations where the need cannot be met by existing housing stock. This support reflected the view that the nature and character of rural housing in the Gower was already suitable to the class fractions singled out by the circular.

The inspector's assessment

The actual matter to be decided by the inspector at the inquiry was the wording of the Gower Villages policies GV3–8 of the local plan, but he considered it valid and proper to take account of the form, content and intended effect of the Secretary of State's direction, which caused the policies to be changed. Once again, therefore, the spotlight fell on circular 30/86 as judgements were made about the legality of the revised policies, and the questions of whether they were only required when needs could not be met through the existing housing stock.

With regard to legality, the inspector ruled on the relevance of Lord Scarman's judgement on the Great Portland Estates, plc *v.* Westminister case thus:

> In my opinion the matter before me sits squarely within the above judgement by the House of Lords. I heard no evidence to suggest that the provision of executive housing under the requirements of Circular 30/86 would be other than exceptional, both in the formulation of development plan policies and proposals, and the operation of development control. As such, reference to such provision should not form part of general policy but rather should be considered as exceptions to it. I do not consider that this would in any way dilute the aims of either Circular 30/86 or the Section 14 Direction of 19th July 1988.
>
> (para. 13.10)

With regard to the question of the need for *new* executive housing in the Gower, the inspector's ruling was equally clear:

the provision of housing does not necessarily mean new housing (although in some cases it could do so). It follows, therefore, that if an adequate supply of existing and suitable housing is available, then no new provision is needed. I conclude on this aspect of the matter that the reference to 'a need not capable of being met from existing stock' in the Mark Three policies is fair, reasonable and relevant.

(*Ibid.* para. 13.11)

Circular 44/87, which superceded circular 30/86, has now been withdrawn.

Conclusions

This examination of the housing–planning relationship in the Gower suggests that there is merit in the study of housing as 'channels of entry' at this scale in a number of comparative localities. The Gower has distinct characteristics. Although an AONB, visually perceived as countryside and possessing some self-advertising 'communities', it is functionally both an extension of the suburbs of Swansea and a holiday resort, and so in many ways it would be foolish to suggest that rurality has much to do with what goes on there. Nevertheless, if it is perceived as rural by potential colonizing fractions, with all of the facets of whatever rural idyll they hold dear, then in one important sense its 'rurality' (as perceived) does make it a target for social recomposition, particularly given the proximity of labour markets and city-based cultural facilities in Swansea. The Gower is also distinct because of the local political representation, which with Labour-controlled local authorities and a Labour MP is somewhat unusual among target areas for colonization by different fractions of the middle class.

What is evident from this kind of study is that the mediation between central and local States and development capital produces a range of different planning arenas in areas even as small as the Gower. Each of these arenas has different housing-market characteristics and therefore each offers different channels of entry to particular fractions of the middle class, thereby making an important contribution to the impacts of local socio-cultural recomposition. What is equally evident is that there is considerable diversity within each planning arena. Some settlements, or parts of settlements, will be much more prestigious and valued than others. Colonization, then, must be viewed flexibly as occurring at different scales ranging from wide expanses of some districts in southern England to smaller, more localized arenas in areas like the Gower. Indeed, it is possible to suggest simultaneous colonization processes by class fractions using different channels of entry occurring in quite close proximity in as varied an area of housing provision as represented by the Gower and the Gower Fringe settlements.

At first glance, the Welsh Office circular 30/86 policies might be interpreted as an overt attempt by the State to ease these processes of middle-class colonization. On deeper reflection, however, it is difficult to compare the senior managers and senior technical staff of potentially inmigrant industry in Wales with the causal powers and relations assumed by service-class fractions in southern England. The one group are sought after by the area concerned, and will be in small enough numbers to be pioneer representatives of their class fraction in a particular locality; the other are seeking out the right kind of places (both personal and positional) to colonize and are sufficiently numerous and credentialled to have a secure channel of entry for their colonization. This is not to say that there are no channels of entry for middle-class fractions in the Gower, nor that different fractions of the middle class are not settling and have not already settled in the Gower. Rather, it appears that circular 30/86 was not specifically designed to aid existing colonizers in their desire to move into selected places in the Gower. Instead it appears to have been designed to create an entirely new channel of entry for a very specific middle-class fraction – senior managers for externally based private-sector firms.

This policy raises interesting questions concerning the relationship between State actions and particular social interests. Circular 30/86 was supported by the Welsh Office and by Welsh business interests as part of the broader promotional campaign aimed at raising the status of the economy of South Wales, and by housebuilders whose vested interest is evident. It was opposed, certainly in the Gower, by all other interests: the colonizing middle-class fractions did not need it because existing channels of entry were suitable for their needs; the previous colonizers who were already in place in Gower villages did not want it, because additional housing would affect their investment and positional goods; and the Labour-controlled local government did not support it, as to them it represented discrimination in the wrong direction – in favour of outside monied classes rather than local low-income classes. The circular should therefore perhaps be seen as a specific realization of a central government's strategy for localized economic regeneration through social restructuring for very specific, and largely 'foot-loose' capital. The precise aims of this strategy remain unclear, perhaps reflecting the way the direction of social restructuring by the central State has become 'a principal which economic theory cannot justify' (Mohan, 1989, p. 82). In the case of the Gower, circular 30/86 appears to mark an attempt to create a new channel of entry for a very specific class fraction. The need for this channel has not been demonstrated, even for the form of social restructuring (for externally controlled 'commercial and industrial enterprises' – circular 30/86, para. 1) favoured by the Welsh Office. Indeed, the Welsh Office decision to focus their attempt to incorporate the circular 30/86 objectives into specific policy modifications in the West Glamorgan–Swansea area was seen by some as a thinly veiled selection of Labour-controlled areas in which to 'make a point'.

This detailed analysis of housing and planning relations is useful only as a starting-point for further research. What is needed now is some equally detailed research designed to discover how these various channels of entry have been exploited by different fractions of old and new middle classes. Only when these two aspects are brought together will the relevance of housing as channels of entry be seen in the socio-cultural recomposition of rural areas.

Notes

1. For a definition of class fractions, see Cloke and Thrift (1990).
2. For a full discussion of the concept of service class, see Cloke and Thrift (1987).

4

STILL NO HOMES FOR LOCALS?
AFFORDABLE HOUSING AND PLANNING
CONTROLS IN RURAL AREAS
Mark Shucksmith

Housing in the countryside is a subject of concern for politicians and the public alike at the present, both because of the expected increase in households necessitating perhaps 2 million additional houses by the year 2001, and also because of the vocal lobbies seeking 'village homes for village people' or seeking to prevent new development in the countryside. These concerns are now widespread, but in areas of great landscape attractiveness the conflict of public policy objectives between countryside protection and the release of land for housing has been particularly acute for some years, and attempts have been made to protect the landscape while remaining responsive to the housing needs of local people. Some thoughts on how this might be achieved were outlined in a speech and a discussion paper by the Secretary of State for the Environment in 1988 (DoE, 1988a, 1988b). Earlier, an innovatory and still controversial policy was implemented in the Lake District by the local planning authority, in an attempt to resolve this conflict of public policy objectives and, at the same time, to influence distributional outcomes in favour of one group (young locals).

This policy response has already been analysed in detail (Shucksmith, 1981; Clark, 1982b), but there are several reasons for re-examining it. First, the empirical basis of the earlier work was somewhat inconclusive, since too little time had elapsed since the introduction of the policy to assess its effects on observed house prices: more data are now available. Second, some criticisms of the earlier work have appeared (Loughlin, 1984; Capstick, 1987) and these may be reviewed. Third, the current government policy initiative to encourage building in villages by housing associations appears to use similar instruments to those criticized as counter-productive and inequitable in the Lake District experiment (Shucksmith, 1981), and it is therefore very relevant at the present time to examine the degree to which such criticisms might also apply to the current initiative.

This chapter therefore re-examines the role planning controls may play in helping or hindering the provision of affordable housing in rural areas, drawing both on the experience of the Lake District National Park and on the more recent initiative of the National Agricultural Centre Rural Trust (NACRT). The emphasis throughout is on the distributional outcomes of policy.

Changing tenure patterns and the incidence of housing disadvantage

In rural areas, the most significant dimension of the housing market is that of tenure. The mechanisms for access to owner-occupation are quite different from those for access to rented accommodation, whether public or private sector. In the Lake District, over half the permanent households are owner-occupiers, and this proportion increased from 52 to 58 per cent during 1971–81, in line with the national trend. The private rented sector declined sharply from 34 to 25 per cent, while the small public sector remained at a fairly constant level (17 per cent) prior to the effects of council-house sales post-1980, which led to a decline in the councils' stock in the national park from 2,264 houses to 1,826 in 1986 (a net loss of 20 per cent). This account of the tenurial structure of the Lake District is complicated by the presence of holiday cottages and second homes. The 1981 Census revealed that 16.4 per cent of dwellings in the Lake District were second homes or holiday cottages, with the proportion rising to over 35 per cent in particularly popular parishes such as Langdale and Patterdale.

Concern about rural housing markets is often couched in terms of young people being forced to leave the area, and an ageing of the population. Overall, the national park's total population has changed very little in the last thirty years. However, this stable aggregate conceals substantial changes in the composition of the national park's population, with a declining number of children and an increasing number of retired people. In 1981, for example, the ratio of retired people to population of working age was 41:100 in the Lake District relative to a national average of 29:100, and the number of persons aged over 60 increased by 10.5 per cent between 1971 and 1981, largely due to retirement migration to the larger and more accessible settlements (Capstick, 1987).

Second-home buyers have generally been blamed for displacing young people in the Lake District and elsewhere. Shucksmith (1981) suggested that a more potent source of external demand for housing in the Lake District than the well-publicized second-home buyer was the retirement migrant, often with a greater ability to pay following the sale of his or her first house. These findings are borne out not only by these demographic trends but by a more recent survey of estate agents, in which 'experienced agents in all parts of the Park were interviewed' (Capstick, 1987, p. 63). This found that 'the second-home market still exists but is less important than it was' and that 'demand from outside the Park is mainly from people moving into the area for retirement'. 'Total outside demand varies, but in popular and accessible parishes, . . . it may reach 50% to 60%' (*ibid*. p. 64). This estimate was confirmed by Capstick's analysis of the origin of solicitors' searches in the Land Registries in 1985. The survey also confirmed that those local people who were able to buy houses in the area were not first-time buyers but middle-aged people with high incomes. Further evidence on this point comes from a survey of the occupiers of several housing estates in the Lake District (LDSPB, 1980), which showed quite conclusively the existence of direct competition between locals and holiday-home and retire-ment-home buyers. The results of this survey are summarized in Table 4.1. Certain caveats apply, however, since the LDSPB admitted at the examination in public of the Structure Plan in 1980 that this sample was drawn selectively to illustrate the extent of such competition: it is not a random sample.

This survey demonstrates that competition for housing exists between local buyers and holiday- and retirement-home buyers at both ends of the housing market within major settlements. The traditional terraces represent the cheapest housing available for purchase in the national park and yet their pattern of occupancy was little different from that of the newer developments. The LDSPB concluded that there was no type of private housing development that would meet local demand without competition from holiday and retirement buyers (*ibid.* Annex 1).

A full analysis of patterns of housing disadvantage in the Lake District is not possible in this chapter, but a more detailed account may be found in Shucksmith (1990). In summary, however, it can be concluded that the more prosperous sections of society have the advantage of access to owner-occupation, and that private rented housing is now only accessible to workers in occupations that offer tied accommodation. Of the poorer groups that remain, the pattern of housing disadvantage depends to a large extent on the processes of public-sector housing provision and allocation. In the Lake District, these processes favour families who seek housing in the main towns, while discriminating against young single people in particular, and against those who wish to remain in their village. The shortage of smaller council houses tends to disadvantage elderly applicants somewhat. While Capstick (1987) seems to confirm this pattern of disadvantage, further analysis is required before a full account of housing disadvantage can be given. One conclusion she reaches, however, is that it will take six to ten years to house those already on councils' priority waiting-lists, and that these lists under-estimate needs.

Table 4.1 Occupiers of selected housing estates

	Ten recent housing developments	Eleven pre-1900 stone-built terraces
Holiday cottage/second home	26%	23%
Retirement home	40%	43%
Local occupier	34%	34%
Total	100%	100%
Total houses in sample	344	100

Source: Calculated from data in LDSPB, 1980.

The policies of local housing authorities

Since the district councils are the local housing authorities for the Lake District area, with responsibility for meeting housing needs, they must be held prima facie responsible for the housing difficulties faced by certain groups. Bennett (1977, p. 28) notes the low provision of council houses in the national park and indicts the local authorities for their lack of imagination and political will. The Cumbria Countryside Conference (1979) explains the lack of council building in terms of higher rural building costs and a failure to observe and to react to hidden needs.

However, the explanation may have as much to do with the historical legacy of a neglect of council-house building in the inter-war period and with financial constraints imposed upon local housing authorities by central government since reorganization in 1974. The proportion of council houses in the national park in 1951, after all, was only 2 per cent (Clark 1982b, p. 63), compared to a national percentage of 18 per cent. Between 1951 and 1976, despite all the factors militating against rural council-house provision, 40 per cent of all new houses built in the national park were in the public sector (LDSPB, 1978, p. 151), although Shucksmith (1981) has shown that during the last seven years of the old authorities (1967–73) only 22 per cent of new houses were council houses – a similar proportion to that pertaining since 1976. This suggests that the low level of council housing in the national park today derives from failings prior to 1951 and after 1969.

More recent failings can be gauged from considering the period 1977–84, during which 87 council houses and 95 housing-association houses were built in the national park out of a total of around 1,000 new houses: together these amount to only about 20 per cent of new houses. The extent to which the post-1974 housing authorities are themselves culpable for this dearth of social housing provision is limited, however, since they have been prevented from building council houses by the financial restrictions progressively imposed by central government since 1975 (*ibid.*). Indeed, capital allocations have been cut by 54 per cent during 1981–7. Capstick (1987, p. 119) is confident that 'had finance been available, [the housing authorities] could well by the present time have reduced somewhat the length of the waiting period for rented housing in the Lake District'. Instead, the stock has diminished as council-house sales have exceeded completions, with a net loss of stock of 20 per cent between 1980 and 1986, despite high prices (even after discount) discouraging sales in the most attractive areas.

The shortage of local-authority housing is certainly a central element in the development of current housing problems facing disadvantaged groups. In terms of policy, while the local housing authorities remained either hamstrung by central-government constraints or preoccupied by housing needs outside the park, depending on one's point of view, the planning authority was confronted by the central dilemma identified above. Should planning controls be tightened in the interests of landscape protection as the limits of acceptable development were approached; or should planning controls be loosened in the interests of relieving the pressure on the housing market and thus, perhaps, helping disadvantaged groups find accommodation inside the national park? It is to the policy response of the LDSPB to this apparently irresolvable dilemma that the discussion now turns.

The Lake District Special Planning Board's housing policies

Since the board's inception in 1974, it has been extremely concerned about housing for local people within the national park. The motive force behind this concern appears to have been the board's realization 'that the difficulties which young people found in obtaining housing were the chief concern of the inhabitants of the Lake District' (Capstick, 1987, p. 133). While the board's primary aim is to preserve and enhance the landscape, it is arguable that the board derives its authority and legitimation not only from the support of powerful and articulate preservationist pressure groups, such as the Friends of the Lake District (Brotherton, 1981), but also from the more tacit support of the people living in the national park.

In 1977 the board therefore announced a new policy to 'restrict completely all new development to that which can be shown to satisfy a local need' 'The villages and towns of the National Park could not be permitted to expand for ever unless the farming industry and the landscape were to suffer irreparably and in many areas the amount of land suitable for housing was beginning to dry up' (LDSPB, 1977b). The board therefore asked any applicant for a residential planning consent to sign an agreement under section 52 of the Town and Country Planning Act 1971, limiting the future occupancy of the house to people employed or about to be employed locally, or retired from local employment. The origins and evolution of the 'homes for locals' policy, or 'section 52 policy', as it became known, are described in Shucksmith (1981).

This policy was identified by the Secretary of State for the Environment as a matter for debate at the examination in public of the Structure Plan in September 1980, where the policy was closely scrutinized. The panel's report and the Secretary of State's proposed modifications were published in August 1981. The panel acknowledged the policy dilemma confronting the LDSPB, but could offer no solution. However, they considered that the board's policies should 'at least be tried' (DoE, 1981b, p. 17). Despite this qualified support, the Secretary of State deleted the policy from the Structure Plan because, in his view, 'it is not in general desirable to seek through planning restrictions to control the disposal of private houses' (DoE, 1983, p. 8) and, moreover, he considered the effect of the policy might be counter-productive.

The effects of this policy, which was implemented from 1977 to 1984 when it was amended in the light of the Secretary of State's deletion of the policy from the Structure Plan, were the subject of detailed analysis in Shucksmith (1981). That analysis is now briefly summarized to allow discussion of more recent empirical information on Lake District house prices during the period of the policy's implementation and subsequent critiques.

The effects of the section-52 policy on the housing market

The aim of the policy was to alleviate the problems of (young) local people in finding suitable accommodation to purchase. By excluding commuters, retirement-home buyers and second-home purchasers from the market for new housing, demand was to some extent diverted from the market for new housing to the (uncontrolled) market for existing housing. Outsiders seeking to purchase homes in the national park were most unlikely to have been deterred by the new policy, but instead tended to compete with locals for the existing stock, over which the LDSPB had no control. Because of this diversion of demand from new housing developments to the stock of houses already existing, it is helpful to consider separately the effects of the policy on the markets for existing houses and for new houses.

The effect of the policy on the market for new housing will have been substantial. The exclusion of non-locals from this market will have caused a marked contraction in demand, since most new houses appear previously to have been bought by non-locals. In addition, builders may have ceased speculative residential developments, partly because of the uncertainties raised by the new policy but principally because of the greater difficulty of acquiring suitable building land with planning permission. Because only smaller developments were to be permitted, with consequently higher land prices, production costs would have been raised (Clark, 1982b, p. 104).

These forces will have tended to offset one another in terms of the effect on price. Further, the supply of new housing is likely to have been sensitive to changes in expected prices (comparatively price elastic), after a time-lag. One would therefore expect there to have been relatively little reduction in the price of new housing, but the policy would be expected to have caused a large decline in the quantity of new housing bought and sold. This is consistent with objectives of restricting development and of retaining for local people the few remaining sites on which housebuilding is acceptable; but it is not likely to have contributed towards meeting the needs of local young people for low-cost housing, which was the *raison d'être* of the policy, according to Capstick (1987, p. 136). Indeed, Capstick (*ibid*. p. 140) has confirmed that 'most of the houses built were not of the type which would have helped local first-time buyers to obtain homes', and therefore the policy 'did more to assist existing home owners . . . than it did to assist new home owners' (*ibid*. p. 144). As non-locals transferred their attention from new houses to the existing housing stock there will have been an increase in the demand for existing housing (a shift of the demand curve). The result of this increased demand is that there will have been an increase in the price of the existing housing stock, as more buyers sought these houses.

The size of the price increase depends upon the sensitivity of supply to changes in price (the price elasticity of supply), and normally one would expect the supply of houses already built and occupied to be highly insensitive to changes in price. The flow of existing dwellings on to the market depends less on price changes than on households' mobility, since households tend to move in response to change in their family or employment circumstances, or to a change in income, and tend not to move in response to changes in house prices. The supply of existing dwellings is therefore likely to be very price inelastic in the medium or short term, and so one would expect the price of existing houses in the Lake District to have risen substantially on a once-for-all basis as a result of the increase in demand for existing houses.

The effects on local people and disadvantaged groups

The effects on local people can be summarized as follows. Local people who could afford to buy new housing will have found prices roughly the same as before, once the shifts in the demand and supply schedules had worked through. The Structure Plan envisaged that new housing would be restricted to infill developments of a type and density appropriate to the character of the settlement, and these were bound to be expensive. Any local person who could afford to buy a new house was unlikely to have been experiencing housing disadvantage in the first place, and could certainly have found housing without the help of this policy. For the disadvantaged groups among permanent residents it was the effect upon the rented sector and upon the price of the cheaper existing housing stock that was relevant.

The effect of the LDSPB's policies on the price of cheaper housing will have been to disadvantage further the less wealthy of the local population in their attempts to become home-owners. Surprisingly, Capstick (1987, p. 136) has argued that this effect is unimportant: 'The Board's whole case . . . was that local young people could not compete in the existing housing market. It was of no consequence to them that further outside buyers would now enter the market'. If this is accepted, then the board's policies will merely have been irrelevant, rather than harmful, to the needs of such groups. However, it seems inevitable that one consequence of a rise in the price of existing houses will have been to exclude marginal local purchasers and thus to add to the numbers of those who cannot afford house purchase and who must look instead to the rented sector.

One might therefore expect there to be more people seeking rented housing. The consequences for the supply of rented accommodation are less clear: while the rise in the vacant-possession price of existing houses, relative to private-sector rents, would have added some further disincentive to private landlords to relet when a protected tenancy ends, the disincentive is so large already that the policy is unlikely to have had any significant effect. In the public sector the main effect is on the cost of land for new building, and it might be expected that local authorities and housing associations would have been able (all other things being equal) to have purchased land subject to section-52 agreements more cheaply than otherwise. However, if this was the real object of the policy, land costs could have been reduced still further for public-sector housebuilding by limiting the definition of local needs to public-sector development only. This course is considered further below in the context of NACRT's initiative.

Empirical corroboration – house prices, 1970–86

This assessment of the effects of the LDSPB's policy has relied purely on deductive analysis rather than on empirical observation. Having deduced these expected consequences, we gathered empirical information on house prices before and after the policy's implementation, in order to check whether the effects deduced above are consistent with observed price trends. The principal source is a data-set of house prices from 1970, gathered from local newspapers. This was the first systematic time-series of house prices in the Lake District to be collected (799 observations to 1980), and this has recently been extended to 1986 (increasing the sample size to 1,261). The details of how this sample was constructed and analysed are given in Shucksmith (1981; 1987). A similar, but far smaller, data-set has recently been collected by Capstick (1987), and this is presented for comparison.

The estimates of house prices are shown in Figure 4.1 (see also Shucksmith, 1987). It should be noted that these are newspaper asking-prices for the most expensive area of the national park, and on both these grounds they may be over-estimates of house prices in the Lake District as a whole.

These results confirm that houses in the south-east Lake District are more expensive, on average, than those elsewhere, reflecting the area's attractiveness.

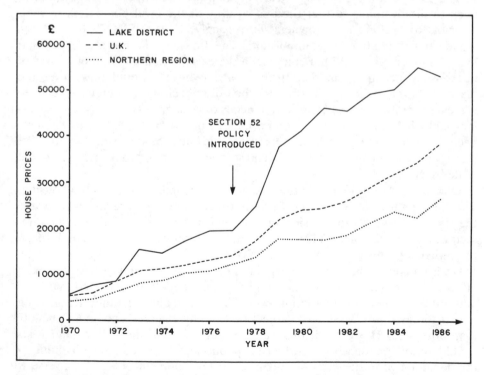

Figure 4.1   Average house prices in the South-east Lake District and regional and national house prices

More revealing, perhaps, are the rates of house-price increase: although Lake District houses are more expensive, for most of the 1970s prices rose little faster than those elsewhere. Between 1970 and 1972, and between 1973 and 1977, house prices in the Lake District increased at a slower rate than national house prices. In the year 1972–3, on the other hand, house prices in the Lake District leapt by 77 per cent, partly due to a boom in second-home buying reinforced by temporarily higher improvement grants in assisted areas.

A period of price stability in the mid-1970s was then followed by sharp price rises from 1978 to 1981, with a notable divergence between local and national house prices in 1978–9 and 1980–1. The total effect of these price increases was that by 1981 houses in the Lake District cost well over twice the 1977 price, whereas UK house prices had increased by only 73 per cent over this period.

Since 1981, Lake District prices have not risen as fast as national house prices in percentage terms, although in absolute terms a differential of about £20,000 has been maintained (ignoring what may be an aberrant price estimate for 1986). Leaving aside that year, it may be seen that Lake District prices since 1973 have maintained a fairly constant differential over UK prices, apart from the period 1977–81, during which the differential of about £5,000 (which characterized the mid-1970s) jumped to the £20,000 gap of the 1980s.

The analysis presented earlier predicted that the major consequence of the LDSPB's policy would be a substantial increase in house prices within and around the national park. The board's policy was introduced in autumn 1977. Information from estate agents suggests that prices started to accelerate in the Lake District in November 1977; and the analysis of house prices presented here shows that the greatest increases appear to have taken place during 1978–9, in line with prior expectations of a lagged response to the Board's policy. Although there were outstanding planning permissions in 1977, and various other lags in the system, the policy created at once the expectation of future house-price increases: this may have contributed to the early rise in prices that followed perhaps sooner than might have been expected.

Price estimates by Capstick (1987), shown in Figure 4.2, also tend to confirm the trends suggested above. Capstick's (1987, p. 137) conclusion that 'Lake District prices rose broadly in line with national trends' appears difficult to reconcile with her published price estimates. It is clear from Figure 4.2 that her series of house prices, despite the much smaller sample, also shows a steep increase relative to UK prices between 1978 and 1980, and it follows closely Shucksmith's (1981) estimates of terraced house prices.

The deductive analysis of the effects of the LDSPB's policy suggested that one would expect a substantial increase in house prices to follow as a consequence of the policy. The empirical finding of a large increase in Lake District house prices following the policy's introduction during 1978–81 is certainly consistent with this expectation. There are other factors that might wholly or partly explain the price rise, such as the renewed growth in second-home numbers in Britain from 1976 to 1977, or perhaps an ability of retirement purchasers to buy a home in the Lake District without requiring a mortgage from building societies who at that time were short of funds. Also, the

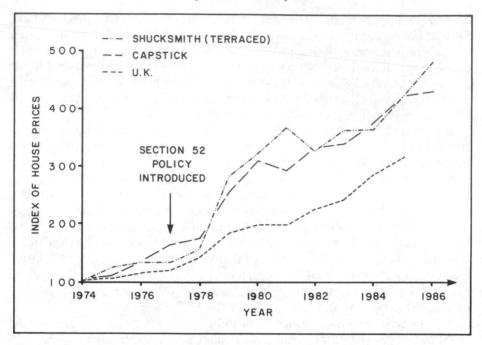

Figure 4.2   Indices of terraced house prices. Note that each sample is drawn from newspaper advertisements. Sample sizes were 4.5 per annum (Capstick, 1987) and 25.3 per annum (Shucksmith, 1981)

expected time-lag before prices rose appears to have been rather shorter than anticipated. In any event, causality cannot be demonstrated merely by observing price trends and this has not been the intention. What has been demonstrated is that Lake District prices tended to rise faster than national prices after the policy's introduction, and this tends, if anything, to corroborate rather than to contradict the a priori analysis. It is therefore consistent with the conclusion that the board's policy led to higher house prices than would have been the case had the more relaxed planning policies of the early 1970s continued.

Two critiques

The first serious critique of these arguments was offered by Loughlin (1984). Loughlin argues that the reduction in residential planning permissions results from the LDSPB's conservationist stance rather than from its section-52 policy. The effects attributed to this policy should, he argues, be blamed instead on the LDSPB's protective attitude towards the landscape (*ibid.* pp. 99–100). Loughlin's logic is quite correct, if approval rates would have declined to the same extent in the absence of the section-52 policy.

But it is not clear that the decline in approval rates would have taken place to the same extent without the existence of section-52 policy; nor is it appropriate to attempt to separate these two policies. The Board's policies arose (Shucksmith, 1981) largely because of a reluctance to impose more restrictive controls on development unless local people could be spared the housing consequences; indeed, Capstick (1987, p. 129) regards the policy 'as a corollary of its strict control of development'. Section-52 policy made possible the board's more restrictive approach because it appeared to allow a more restrictive policy also to help local people find houses. A number of the LDSPB's statements reveal this attitude. For example, 'In the Lake District, however, the board felt that whilst stricter development control was justified it should be matched with a policy which would attempt to help with the particular difficulties of local people' (LDSPB, 1980, p. 1). Clark (1982b, p. 96) also notes an intertwining of the LDSPB's concerns for landscape and housing in his account of the policy's origins. Similarly, Capstick (1987) explains how the policy arose from public reaction to the board's preliminary intimations that it would adopt a more restrictive attitude to new housing developments. To retain the legitimation and the support of the park's population, the board was unwilling merely to impose such restrictions without addressing local housing needs.

Loughlin is therefore incorrect to attempt to separate the local user policy from the board's conservationist stance: the policy both follows from the more restrictive stance and at the same time legitimizes it, in that it is unlikely that the board would have felt able to adopt such a restrictively conservationist stance without demonstrating (both to the people of the park and to themselves) that the housing interests of local people would be protected. The two elements are inseparable.

More recently, Capstick (1987) has offered a critique of the economic analysis above. She disputes the conclusion that the policy's effect will include a rise in the price of existing houses, arguing instead (p. 143) that the price would remain unchanged because 'demand would transfer equally to the older stock from incomers, and to the new stock from residents'. This argument, however, is fallacious, resting on an implicit and untenable assumption that the supply of new housing is infinitely price inelastic (i.e. that builders will construct a fixed number of houses whatever the selling price). Indeed, this is an extraordinary implicit assumption for an economist to make.

If the supply of new houses is at all responsive to price changes, on the other hand, then the quantity supplied and consumed will fall and consequently demand will not transfer equally to the new stock from residents. Indeed, if as is likely the supply of new houses is price elastic, then there will be relatively few additional residents transferring from the existing stock to new houses, and the net effect will be an increase in the demand for existing houses as originally suggested.

Can planning help? The NACRT initiative

According to the LDSPB (1979), several other planning authorities were monitoring the progress of the 'locals only' policy with a view to implementing it themselves if it proved effective. This proliferation was prevented by the

Secretary of State's opposition to the policy. However, now that the government has endorsed the use of such planning conditions as part of its rural housing strategy (*Hansard*, 3 February 1989), and accepted that housing provision for local people may properly be a material planning consideration, it is important to be clear about the particular shortcomings of the Lake District policy.

Policy mechanisms employed to discriminate in favour of disadvantaged groups, such as young local people, require two attributes if they are to be successful. Houses must be allocated by non-market mechanisms if they are not to be allocated simply according to market strength; and the houses must be affordable by the disadvantaged groups intended. Otherwise, houses will continue to go to those with the greatest ability to pay, even if supply is restricted.

From this perspective, the weaknesses of the LDSPB's locals only policy were twofold. In the first place, the private land market was not bypassed or replaced but merely sub-divided into two: within each segment (existing houses for all, new houses for locals only) the socially regressive effects of a severe restriction on supply remained. Second, and related to this, the policy attempted to discriminate on too crude a basis between locals and non-locals rather than on the basis of housing disadvantage and need. What was really required was an alternative allocative mechanism that, unlike the private markets, favoured either those groups with the greatest housing needs or those groups with the most to contribute to national-park purposes. In addition, a subsidy would probably be required to make houses affordable to such groups.

There are, then, two necessary elements (Shucksmith and Watkins, 1988): the building of low-cost housing in rural areas, affordable by those with the least means; and the allocation of that housing such that it goes to meet the needs of those households. Neither of these elements was present in the Lake District experiment.

In contrast, these elements are both present in the initiative pioneered by the National Agricultural Centre Rural Trust (NACRT), which has been responsible for the government's change of attitude towards local-needs conditions. NACRT promotes the building of low-cost housing in English villages by rural housing associations. According to Constable (1988), land costs are a major obstacle to such provision, typically accounting for about 40 per cent of the total cost of provision if the open market value is paid. In order to reduce costs, and to pass the saving on to the eventual consumers, NACRT has developed a method of acquiring land cheaply, at a price well below its value with residential planning consent. NACRT seeks sites peripheral to the village, not zoned for housing in local plans, and hence worth substantially less than full development value. If the landowner is willing to sell the land to NACRT, and the village community approves, NACRT approaches the local planning authority, asking it to make an exception to its normal development control policy and to give permission for the rural housing association to develop the site for low-cost housing for specified groups. Frequently a section-52 agreement is used to define these terms, and effectively to restrict the planning consent to the provision of social housing.

In this application, then, the section-52 agreement is no longer a crude allocative device, subject to the criticisms above: instead, it is a means both of reducing land costs (so assisting the provision of affordable housing) and of defining the tenure of the houses that will be built (wholly or partly rented from housing associations, who can exercise more precise control ensuring their allocation according to need). The result is provision of low-cost housing for those in need.

While this *ad hoc* arrangement appears to work well at present, it works only at a very small scale of provision. Only a few houses each year have been built under NACRT's initiative throughout rural England: 305 were due to be built in 1988–9, and the latest enhanced programme announced by the Minister (November, 1989) will aim for a target of 1,250 houses in 1990–1, rising to 1,500 (1991–2) and 1,850 (1992–3) throughout rural England. This compares with a recent estimate of need of 370,000 (Clark, 1990). If this initiative is to be the basis of a strategy for broadening access to rural housing on the scale required to meet needs, three crucial concerns arise. First, will landowners release land on the scale and at the price required? Second, will the Housing Corporation allow rural housing associations enough funds to build on the scale required? And third, will planners everywhere decide that meeting local housing needs justifies the 'sacrifice of normal planning considerations (Greenwood, 1989)?'

At the larger scale of operations, which would be appropriate to the evidence of needs, it will not be sufficient to rely on time-consuming searches and the goodwill of a few philanthropic landowners. Furthermore, as the scale of activity grows, planners will become less willing to make exceptions on greenfield sites not intended for building, because of the fear of precedent and the cumulative impact on the landscape. More landowners are likely to hold back, in anticipation of full development gains. A more general solution will be required, which must be built into planning policies and land values rather than tacked on as an occasional exception to policy. Nevertheless, it should build on the successful NACRT approach, which offers a mechanism for the provision of affordable housing for lower-income groups, rather than on that of the Lake District experiment.

The essence of the NACRT approach is concerned with reducing the development value of the land, and hence the land cost to the developer, and with passing on that reduction to consumers of limited means. This could be achieved through the creation of a new planning-use class, through which some land would be earmarked for social housing schemes only. The effect of this would be to restrict the bidders for such land to housing associations and local authorities, and to place them in a position of monopoly. Landowners of land zoned in this way would be faced with a choice: either to accept a low price (perhaps ten times farmland prices) offered by the housing association or to hold the land in the hope that a later development-plan review or a planning appeal would confer on them the full development value. Balancing these probabilities, at some price above farmland value but well below the full development value, landowners would be likely to sell to housing associations. The forging of a consensus between local and central government over the planning status of such land would be essential to the success of such an approach. If landowners' expectations of development gains were effectively dampened on land zoned for social housing in this way, then rural housing associations could expect to purchase land much more cheaply and on a scale sufficient to address rural housing needs.

Such an approach also relies on the planning authority zoning sufficient land for social housing in order for the programme to keep pace with needs. In most areas one might expect that this form of land release would be additional to land release for private development, so having a neutral effect on the private land and housing markets. In areas of particular landscape value (such as the Lake District) land might only be released for social housing, however, and this would necessarily result in rising house prices in the private sector. This would have less serious implications than under the former policy, though, since disadvantaged groups would enjoy access to the social-housing sector.

Conclusion

There is little doubt that planning controls act as a supply-side constraint in much of rural Britain, inflating rural property prices and reinforcing the social exclusivity of the countryside, as demonstrated in Chapters 2 and 3. It does seem that planning controls could also be employed for the benefit of disadvantaged groups, however, through the designation of land upon which social housing alone can be built. The NACRT initiative has used such controls, through the device of section-52 agreements, to allow rural housing associations to acquire cheap land on which to build low-cost homes that are allocated to those in housing need.

It is important, however, that this use of section-52 agreements should not be associated with the Lake District 'locals only' policy of 1977–84. This chapter has reviewed the earlier criticisms made of that policy and has confirmed the failure of that policy to help disadvantaged groups. Two weaknesses in particular have been identified: the failure to promote the provision of low-cost housing, affordable by those groups intended to benefit; and the failure to ensure the allocation to those groups of such houses as were built. Instead, these were allocated by the market according to ability to pay.

Future attempts to cater for disadvantaged groups in rural areas must ensure both the provision of low-cost housing, affordable by those client groups, and its allocation to those groups. While the NACRT initiative meets both these conditions, it succeeds only at a very small scale at present. For this approach to match the scale of needs in rural England, a new use class for social housing is required. Even then the success of such a strategy will depend on the attitudes of landowners, on the zoning of such land by planning authorities with support from central government and on the investment priorities of the Housing Corporation.

5

PEOPLE WORKING IN FARMING: THE CHANGING NATURE OF FARMWORK
Gordon Clark

British farming has changed profoundly, as has agriculture in much of the developed world. The relentless logic of capitalism – the need to make adequate profits to stay in business – eventually induces every farmer to seek higher standards of managerial and technical proficiency. Higher yields and the more productive use of all one's assets are the only long-term ways of continuing in farming. Despite all the criticisms of the levels of farm support, farmers do compete with each other to capture for themselves the best prices. The extensive systems of subsidy that have persisted for half a century in many countries have acted as no more than a partial shelter from the effects of this competition: they have been like a parachute that slows, but does not reverse, farming's declining income.

Now even this parachute is being cut away. Self-sufficiency in food supply and endemic surpluses are creating such budgetary pressures in the European Community, and such political pressure from the USA, that subsidies are being reduced, restricted or withdrawn. In the UK, farmers have been under additional pressures caused by the relatively unfavourable course of the national economy since the Second World War – rapid inflation of input prices and wages, periods of high, real interest rates and slow population growth. When these pressures are combined, it is hardly surprising British farming is being re-organized. This chapter considers aspects of how employment in agriculture has changed: who works in farming, what they do and the terms under which they work. The data for such a study are not ideal, but they are adequate for elucidating the principal trends.[1,2]

Fewer farms, farmers and workers

The total amount of income from food production in the UK more than halved between the mid-1970s and the mid-1980s according to one of the ways of measuring farm income; the terms of trade have moved steadily against farmers (Marks, 1989, pp. 149–51). One consequence of this has been that farming income has been shared among fewer farmers. In 1931 there were around 300,000 farmers in Great Britain as there had been in 1851: by 1987 the number had fallen to about a quarter of a million (Marks, 1989; Ministry of Agriculture, Fisheries and Food *et al.*, 1968). This is a more modest decline (17 per cent) than in the numbers of farms or farmworkers. There was a 49-per-cent drop in the number of agricultural holdings between 1875 and 1985 (a trend exaggerated by changes in statistical practices) as separate holdings were amalgamated into larger units (Clark, 1979; endnote 2). Even more remarkable has been the change in the number of farmworkers. In the 1850s the total agricultural workforce peaked at 2 million people (23 per cent of the working population of Great Britain).

The decline in the number of hired workers was initially steady and, after the Second World War, very rapid indeed, giving a fall of 83 per cent by 1987 (Figure 5.1). The ratio of hired workers to farmers fell from 5.6:1 in 1851 to 2.8:1 in 1931, and to 1.1:1 in 1987. Farming now accounts for only 2.4 per cent of the civilian workforce in the UK. During the 1960s, over a quarter of a million workers left the land, which was the largest decline recorded in any decade: by the 1980s the exodus had slackened to about 5,000 a year. None the less, the 610,000 people (farmers, their families and staff) who worked on British farms in 1987 still represented a larger section of the national workforce than is found in the energy sector, the chemical industry or electrical engineering. Not all of them work full time, but counting the part-time and casual staff proportionately, the total amount of labour applied to British agriculture fell by 13 per cent between 1975 and 1985 (Eurostat, 1989, p. 44). Another change in the last fifty years has been the growing employment in farming's supply industries (e.g. machinery and fertilizer manufacture), in food transport and in the processing and marketing of food. The National Farmers' Union has estimated that 2.1 million people (10 per cent of the workforce) are employed in the food chain – a figure of as much political as economic significance since it includes imported foods (Centre of Management in Agriculture, 1987, p. 50).

Not only are there fewer hired workers but there are fewer farms with any hired labour at all – only 28 per cent of all holdings in England and Wales in 1986 employed any workers compared with almost 40 per cent in 1951 (Ministry of Agriculture, Fisheries and Food *et al.*, 1989). Definitional changes in the Census suggest that these figures are probably under-estimating the decline by about 3 per cent. So three quarters of holdings are truly family farms in the sense that the farmer and family constitute nearly all the workforce. On those holdings where there are still hired employees, the steepest decline has been in holdings employing five or more workers (Marks, 1989, p. 151). So where workers are still found, they tend to be in smaller groups than previously. Half the hired workforce is spread very thinly in ones or twos to a holding. The other half is concentrated on just 6 per cent of all UK holdings. None the less, it is worth remembering that despite this decline in hired workers, they are still twice as important a component of the British farm workforce as in the European Community generally.

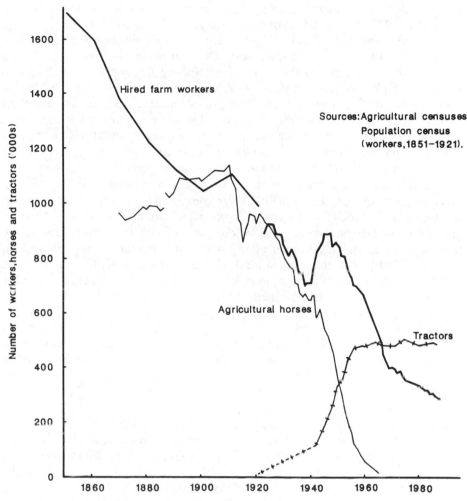

Figure 5.1 *The decline in the number of hired farmworkers 1851–1987 (Source: Agricultural Censuses;* Population Census *(workers, 1851–1921))*

Hired workers

The decline in numbers

The reasons for the decline are not hard to find. Since 1960 the earnings of full-time hired farmworkers have advanced 8 percentage points nearer to parity with national average earnings (Marks, 1989). In the longer run, Marks (*ibid.*) has estimated that they have risen from half the industrial average in the mid-nineteenth century to four fifths now. However, farmworkers are still one of the lowest paid occupational groups (Winyard, 1982). What is interesting is that this real rise in pay has been achieved despite the inability of the farmworkers'

trades union to organize effective industrial action – the last strike was in
Norfolk in 1923 and only a fifth of farmworkers now belong to a union (Bain and
Price, 1983). The Agricultural Wages Board was re-established in England and
Wales in 1924 and, arguably, it has served the remaining workers well. To the
wage costs must be added the expense of housing farmworkers: tied accommodation is not as common as previously, but it is still received by nearly half the
full-time workers. In areas of high house prices – around cities or in scenic parts
of southern England particularly – considerable income is forgone by not being
able to sell the house to a commuter or retiree. Yet many rural housing markets
would be too expensive for those on farm wages and so tied housing may be the
only way of keeping workers in the locality: the cost falls on the farmers.

The decline in the number of farmworkers has been matched by the
mechanization of agriculture. In 1911 there were as many farmworkers as horses
on farms – 1.1 million of each (Figure 5.1). Even as late as 1946 over
half-a-million horses and 889,000 workers were left, the latter boosted by
prisoners-of-war and the Women's Land Army. By 1965 there were so few farm
horses (21,000) that the Agricultural Census stopped counting them. The
steepest decline in the number of horses was about a decade earlier than that for
the workers, and it coincided exactly with the advance of the tractor.

The mechanization of agriculture cannot be said to have started in any
particular year, of course. Steam ploughing was introduced in the nineteenth
century, the reaper-binder was adopted on farms after 1879 and the government
sponsored the use of 600 steam ploughs during the First World War to combat
labour shortages. The 1920s saw several major technical improvements to the
tractor, which increased its reliability, power and quality of work. The 12,000 or
so tractors in 1921 had multiplied tenfold by 1942 and fortyfold by the early
1960s (Marks, 1989, p. 146). Since then there has been only a fractional change
in their numbers but a doubling of their working capacity as more powerful
models were introduced. A similar story can be seen for the combine harvester.
There were about fifty in Great Britain in 1934 (compared with 61,000 in the
USA in 1930) and 67,000 in the mid-1960s, slipping back to 54,000 more
powerful machines in the mid-1980s. Equipment has substituted capital for
labour, and energy from electricity or oil for manpower and real horse-power.
Each remaining worker can plough faster, harvest crops more quickly or look
after more animals. Such extra labour productivity has easily outstripped the
physical growth of farms and has allowed farmers to get by with fewer staff.
Despite higher real wages, farmers have kept their proportion of spending on
hired labour almost constant since the mid-1950s by means of mechanization and
fewer staff (*ibid.* p. 151) and it has actually fallen in real terms by a fifth since
1945 (*ibid.* p. 17).

So far the discussion of the shrinking farm labour force might have given the
impression that workers were being made redundant because of mechanization.
In a general sense that is true for agriculture as a whole, but for individual
workers there was a wide variety of circumstances surrounding their leaving the
farm. Drudy (1978) and Drudy and Drudy (1979) studied workers leaving
farming in north Norfolk in the 1960s when the contraction was at its most rapid;

while a third were made redundant, a quarter left of their own accord because of the low wages. McIntosh's (1972) study in Scotland in the late 1960s suggested that 70 per cent of workers left voluntarily, the younger ones mostly attracted by better pay and training in the towns, and the older ones more likely to move for better pay or promotion within farming. The suggestion is therefore that the youngest workers and the newest recruits are attracted out of farming by better prospects elsewhere, whereas the older workers who leave farming before retirement are more likely to be forced out of the industry against their will. For individual farmers, mechanization was at least as much a reaction to labour shortages as it was the cause of the shrinking farm labour force.

It must also be remembered that behind the figures for the net loss of farmworkers there lie much larger flows of workers into and out of the industry. Wagstaff (1971; 1972) showed that these gross flows were double the size of the net change, reinforcing the point that the farm labour force should not be thought of as a diminishing corps of farmworkers but rather as a reservoir with workers flowing out rather faster than they are being fed in. The Census simply records the number employed each 4 June and so conceals this dynamism. Only 4 per cent of those employed on Scottish farms in 1950 were still in agriculture twenty years later. Farmwork is not a career for most staff but a phase in their lives. Traditionally it comes soon after leaving school – few take their first job in farming after they are 25 – and it is a brief period since four fifths of youths in farmwork will have left agriculture before they are 25. For the remainder, the longer they stay in farming, the less likely they are to leave of their own accord before their retirement (*ibid.*). This makes the full-time farm workforce a very young group indeed, with two thirds under 36 years of age (Burrell, Hill and Midland, 1984, p. 40). Martin (1977) has suggested that the advent of machinery and formalized training is limiting the 'wastage' of younger workers, but his evidence is not strong.

Conditions of farmwork

Technical progress in farming has obviously been immense in terms of machinery and chemicals, but the demands made of workers have also been changing. They need more technical skills, they must be more self-disciplined (since they are often working alone and unsupervised) and they need to be aware of the paramount need for precision in their work. This affects animal health, the efficiency of resource use, the effectiveness of chemicals and the germination of crops, not to mention their own safety and that of the public. They also need to be able to work hard in concentrated bursts of activity. The speed of modern machinery allows the farmer to catch up on tasks if conditions have been unfavourable for ploughing or harvesting, for example, but this places additional strains on the workers different from the sheer physical drudgery that was formerly the worker's lot.

The growing importance of more skilled workers is demonstrated by how the structure of agricultural wages has evolved. There has been a system of setting statutory minimum agricultural wages continuously since 1924 (in Scotland, since 1937). These wages show a widening of the differential between general farmworkers and their more skilled colleagues. Between 1972 and 1985 the minimum wage for craft-workers in England and Wales rose by 23 per cent more than that for ordinary workers. In Scotland, dairy stockmen's basic wages pulled ahead of the general workers' by 38 per cent between 1965–6 and 1987–8. One reaction to this wage differential could be a de-skilling of the workforce with more of the cheaper grades being used. There is some evidence for this but it is not conclusive. A study of full-time workers' earnings in 1981 (Philpott and Tyler, 1987) showed that foremen and dairy cowmen earned much above the average as did workers on large farms. One in eight workers earned over £200 a week in 1987 and the same proportion earned under £110 a week, these being particularly in horticulture.

Skills have to be learnt, of course, whether they are for specialists or general workers. An important development here has been the expansion of the Agricultural Training Board (ATB) and its local committees, which have been running short agricultural courses since 1966 to supplement the higher-level work of the universities and agricultural colleges. Agricultural training is not new; rather it has been increased in scale and become more formalized: there were three times as many ATB courses in 1985 as four years earlier, and 8,000 people participated in them – the equivalent of 6.6 per cent of the full-time hired workforce (ATB, 1985, and unpublished data). Instruction 'on the job' has increasingly been replaced by formal off-farm courses providing what local farmers perceive their workers to need.

Full-time farmworkers in England and Wales work longer hours than most other employees – on average, four hours a week more in 1987 compared with employees on full-time rates in manufacturing (*Employment Gazette*, 1988). The basic working week has been 40 hours in England and Wales since 1975 (it was ten hours more in the inter-war years), but an average of seven hours overtime is a normal feature of farmwork, as are seasonal and short-term peaks of activity. Being a farmworker is no more a predictable 9-to-5 weekday job than is being a farmer. Among the attractions of farmwork are the variety of tasks as well as working alone, with some responsibility, in the countryside. However, these very features may contribute to the high accident rate in farming. Fatalities on the farm are usually three or four times more common than in manufacturing though, fortunately, on a downward trend (Goddard, 1988). Injury or occupational diseases, such as 'farmer's lung', are more insidious hazards and they affect both farmers' and workers' careers.

Work patterns

So far the discussion has centred on regular full-time hired employees, but these now account for only 36 per cent of the hired workforce compared with 46 per cent in 1972. Before the 1970s family workers and managers were less fully enumerated in the Agricultural Census, and so long-run comparable data are not available. However, between 1946 and 1966 the proportion of the hired workforce who were full time declined from 83 per cent to 70 per cent (Ministry of Agriculture, Fisheries and Food *et al.*, 1968, p. 62). Today nearly half the hired workers are either part time (19 per cent) or seasonal/casual (30 per cent), compared with 27 per cent in 1960 and 13 per cent in 1938. What has happened is that since 1960 the number of full-time workers has declined nearly twice as fast as the number of part-timers and, indeed, in 1986 there were slightly more seasonal and casual workers than there were in 1955 when they were first enumerated separately in Great Britain.

It is tempting to over-interpret this change in the composition of the workforce – to see it as evidence of an emerging dual economy with increasing numbers of 'secondary' workers with low pay, irregular work and little security. In fact, some important aspects of the changes are unclear – we do not know how many of the seasonal and casual workers are members of the farmer's family. Nor do we know how many have other jobs elsewhere in the economy. This is a serious omission because the various components of the workforce have been treated very differently. Male workers (whether family members or hired, part time or full time) have all been shed at roughly the average rate. Female workers show a much more complex pattern. Female workers who are members of the farmer's family have almost disappeared from the scene – in 1987 there were only a third of the numbers in the early 1970s, with the decline being greater still for full-timers. In the sharpest contrast, female workers who are hired show the smallest decline of any of the employee groups – a third fewer full-time hired women and only 15 per cent fewer part-timers. This suggests that there is now less formal and regular involvement for farmers' wives in manual farmwork than previously, whereas hiring other women for part-time work is of continuing importance. Among part-time staff, women actually outnumber men in England where the large horticultural sector provides many jobs conventionally seen as 'suitable' for women. Among the seasonal and casual staff, almost the same pattern is found: in England the women almost equal the number of men. The low earnings in the horticulture sector are clearly related to the predominance of female workers. We do not know whether the buoyancy in their numbers reflects increasing use of outside workers instead of the family, or the down-grading of part-time jobs into casual ones or the relacement of hired staff by occasional contributions by family members. What is incontrovertible is that the increasing importance of seasonal and casual workers highlights both the falling need for farm labour and the irregularity of the remaining jobs, which precludes regular workers.

Of course, irregular work patterns are not new: what is different is that the peaks now form such a large proportion of the year's work on many farms. In the past, periods of hectic activity were traditionally covered by workers brought in for a short period – there were the contract sheep-shearers, the harvest labourers (even in the nineteenth century – Johnson, 1967), the gangmasters of the eastern counties, the East End hop-pickers and the school children given special harvest holidays. Sometimes the farmer would hire the short-term staff. If he needed a large number or specialist equipment (e.g. a pea viner or aerial spraying), then he might use a contractor or gangmaster to organize the work. The 'peakiness' of farmwork has been accentuated by mechanization because each job can now be done so quickly that it has been finished long before the next task starts. The peakiness has also been intensified by a simplification of farm systems with each farmer concentrating on fewer enterprises. In the past the peaks of work were flatter (i.e. jobs took longer) and there were so many peaks from the range of enterprises on each farm that they merged into a plateau of continuous labour demand. Nowadays the peaks are sharper and fewer so that, despite farm enlargement, they stand out from a plain of limited activity. The opportunity exists to cover the peaks with short-term family or hired staff and for the farmer to do the rest of the work alone. Also of importance is the legislative background, which favours using short-term staff since they are not subject to redundancy legislation, they do not require a tied cottage, and National Insurance and pension costs will be much less onerous.

The contrast has always been between the smaller and larger farmers. The latter can justify their own machinery for their specialist enterprises. They retain the once-traditional workforce of full-time employees and perhaps the gradations of staff such as foremen, craft-workers and general workers (Symes and Marsden, 1984). It is the smaller-scale farmers who are increasingly dependent on themselves and their families for getting everything done – more truly than ever, theirs are family farms in terms of labour as well as management and investment. On the smallest English farms the owner and family workers provide 90 per cent of all the labour, on medium-sized farms 70–80 per cent and on the largest farms about 30 per cent (Furness, 1983). In 1986 farming families accounted for nearly two thirds of all the labour used on British farms (Eurostat, 1987); hence we must now turn our attention to the farmers and their families in their increasingly important role as farmworkers.

The farmer and family

Some 10–20 per cent of farmers are unmarried – elderly batchelors seem to be commonest in the remoter parts of the UK. Estimates of the percentage of women farmers vary between 3 per cent (Gasson, 1989) and 10 per cent, rising to a fifth among the smallest farms run perhaps by widows (Wagstaff, 1970). These groups apart, the farmers' families most often means the farmers' wives and children.

Farmers' wives as workers

The role of farmers' wives is problematic and unclear. Wives engaged in farming have been separately enumerated in the Agricultural Census only since 1977 and the number has held steady at around 77,000, which is about a quarter the number of farmers, partners and directors (some of whom will also be farmers' wives and women as farmers in their own right). This stability is in sharp contrast to the severe decline in the number of female family workers employed full time or part time. Along with daughters and other female relatives, wives are probably being employed less often in a formal sense by their husbands, the farmers. In total, wives probably provide about 5 per cent of the labour input to

British farming, with their contribution being much higher than this on the smaller farms (Gasson, 1989).

Little (1987b) argues that the role of farmers' wives is best seen as a subordinate one within a patriarchic structure. Farms pass to sons, though they may have to buy out daughters' shares. The farmer runs the farm and his wife's role is secondary. Even if her labour input is considerable, it will be treated with the paternalist benevolence Newby (1977) identified as the farmworker's lot. Symes and Marsden (1983) have followed Bouquet's (1985) argument that the growing technical component of modern farming has acted to cut wives off from farmwork – an argument many women would find controversial. The wife's main role is in developing the next generation of farmers and caring for the current generation.

Buchanan, Errington and Giles (1982) conducted a sample survey of mostly large-scale farms in south-central England that showed that the wife's role centred on the farm office (book-keeper and receptionist) even when a secretary was employed. Hastings's survey (1987–8) reinforced the wives' role in business management and financial control and their lesser involvement in manual and technical matters except on small livestock farms (where wives often reared stock) and on smaller horticultural units. A second task was as an additional pair of hands when needed – rarely on larger farms except in a real emergency; more common on smaller farms with no hired staff when she might work frequently and provide cover for her husband when he was ill or away. A third role he identified was a confidante, sounding-board and giver of advice – someone with whom to discuss issues. This last role is the hardest to assess with any accuracy since one cannot know whether the advice was welcomed or taken (Errington, 1983). In some cases the wife might be in a position to exert considerable influence if she has provided her own inherited land or capital for the farm.

What is clear is that these roles focus on the farmhouse (which usually doubles as farm office) reinforcing the home-centred and caring aspects of the feminine stereotype of wife, mother, carer and rearer. Gasson (1981) postulated three types of wife on farms. The 'working farm wife' is heavily involved in running the smaller farms, both manually and managerially. The 'farm housewife' on the larger farms has much less involvement and it is rarely manual. There is also the fairly rare 'woman farmer' who runs the farm on an almost equal footing with her husband. Hastings's survey (1987–8) hinted at how wives believed that their work was important to the farm but felt under-trained and under-valued. Symes and Marsden (1983) suggest that her role is reduced to a supportive one, casualized and dominated by the mundane – in short, more unequal. Wives' careers off the farm will be discussed later. Of course, these role-based approaches are descriptive rather than analytic of the processes leading to and perpetuating this economically peripheral status.

Part-time work and other sources of income

The assumption that farmers and their families are no more than producers of food, and spend all their time, skill and capital on this, has always been misleading (Robson, Gasson and Hill, 1987). They often have other sources of income and other jobs. Measuring the extent of such 'pluriactivity' is not easy, but its scale is certainly considerable. Inland Revenue figures for 1987[3] show that only 11 per cent of farmers are solely dependent on their income from self-employment in agriculture (Ministry of Agriculture, Fisheries and Food, 1987). Three quarters also have some investment income and an eighth some income from another job. Nearly 40 per cent of all the income for farmers comes from outside food production on their own farms – 19 per cent from investments, 15 per cent from other jobs and 6 per cent from pensions.

Behind these national averages there are considerable variations. In 1987 the 2,000 wealthiest farmers (under 1 per cent of the total) earned 16 per cent of all the investment income accruing to British farmers and 12 per cent of the income from other jobs. The poorest farmers (11 per cent of the total) shared only 1.5 per cent of farmers' total income and relied heavily on pensions, whereas the wealthiest quarter earned three fifths of total income. The large amount of agricultural business profits earned by farmers' wives suggests income diverted to them for fiscal purposes rather than a measure of their real involvement in farming. These 'tax-efficient' wives outnumbered considerably those clearly employed separately in their own right.

There are other ways of assessing how many farms are part time. The Agricultural Census classifies as part time those holdings with a labour requirement of under 250 standard man-days: these increased from 49 per cent of all holdings in 1978 to 53 per cent in 1985. Using 'business size units' as the measure of farm size shows the same slow rise in farms unlikely to give a full job and income to a farmer. The European Community's farm-structure survey of 1985 (Eurostat, 1987) revealed that 21 per cent of UK farmers have some 'other gainful activity' – the third lowest percentage in any member state, reflecting the larger size of British farms. But 29 per cent of British farmers' wives have a job – the third highest percentage in the EC. In all cases the older farmers are less likely to have another job, and the smallest-scale farmers (under 1 European Size Unit or 5 hectares) are four to five times more likely to have another job compared with the larger farmers. This size bias is also true of farmers' wives, but not so clearly. Wives on the smallest farms are as likely to have another job as their husbands but the wives on the largest farms are three times more likely to have another job than their husbands.

Again we see a complex interplay of different opportunities and necessities for men and women. Smaller-scale farmers have to supplement their low agricultural incomes as do their wives. Larger-scale farmers do not need another income but may choose to pursue their original non-agricultural job while farming or may diversify out of farming as a way of using spare capital. Their wives are more likely to be able to obtain a job or set up in business using their education, their freedom from farmwork and either their husbands' capital or their own as the basis for a business. They choose to work for pleasure or as an assertion of their independence: they do not need to. The 'other jobs' of the large-scale farmers and their wives yield much more extra income than the jobs held by the smaller-scale farmers whose farming income is in such need of supplementation.

Gasson's (1983a) study of farmers' other jobs showed that these included working for another farmer, self-employment outside farming (e.g. a milk round) and tourist accommodation. It was found that the latter role usually involved the farmer's wife, who exchanged looking after farm servants for caring for tourists. This is a complex process: some farmers are becoming proletarianized as they accept outside employment in the manner of the factory-and-farm workers of West Germany. Others bolster the small-business sector with new self-employed ventures, reinforcing the intrinsic advantages of farming as 'being your own master'. In other cases the wife's role is enhanced as her job (be it teacher or provider of B & B) is financially more important. Generally, farm tourism gives high margins but a small total income, so its importance is often as much symbolic as directly economic (Bouquet, 1985).

Part-time farming is not a new phenomenon, of course. Wagstaff (1970) noted that other jobs were held by half of Scotland's smallest farmers and a third of the slightly larger ones – results similar to those of Scola (1961). A dual economy of small-scale farming and mining was common in the nineteenth century in the Pennines and North Wales, and a similar linkage of farming and a wide variety of jobs is still found in the crofting areas of northern Scotland and the Isles. The National Farm Survey of 1941–3 (MAF, 1946) reported that 26 per cent of farmers had another job. The other job may supplement a meagre farm, while the farm provides a fall-back of income and subsistence when, say, the mine closes, weaving takes a downturn or there is an over-supply of farm accommodation. Pluriactivity can be a way of coping with the risk of volatile incomes as well as boosting low earnings. A distinctive rhythm of work is often found, farming in summer and weaving in winter, or one job weekdays and the other at weekends. The key to multiple job holding is how to arrange one's time to fit in both satisfactorily.

For some farmers the combination of jobs is a passing phase as they ease themselves out of farming or gradually move into agriculture. Gasson (1983b) and Wagstaff (1970) both suggested that these two transitional groups were roughly equal in size in the 1970s. For many on the largest and smallest farms the arrangement is effectively permanent, however. What is often overlooked is that many small-scale farmers have no other job – pensions or poverty are their lot. At the other end of the spectrum are the wealthy who have moved into farming after amassing capital elsewhere in the economy. The urban hobby farmer is most common in the Home Counties, enjoying a landed-gentry status: whether farming is the most efficient use of their wealth is a moot point (Gasson, 1966). Farmers in the South East have the most income from investments (Ministry of Agriculture, Fisheries and Food, 1987). Other wealthy farmers will diversify out of farming often into related industries, such as marketing, contracting or food processing. They may be seeking a way of investing profitably the surplus capital they have accumulated from farming (Shucksmith *et al.*, 1989).

'Part-time farming' is a useful statistical category, but theoretically it is a

other, wealthy agri-business people are controlling extensive assets, dominating production for the food chain and – under certain circumstances – employing managers to use resources and staff as cost-effectively as possible.

Acknowledgements

The author acknowledges the financial support of the ESRC in the research on farm managers (research grant F00232208) and of Dr A. McAuley, the research assistant on the project.

Notes

1. In this chapter the data usually refer to Great Britain unless otherwise stated. Prior to 1922, UK data exclude southern Ireland. Wherever possible the most recent data have been used – usually for 1987 (Ministry of Agriculture, Fisheries and Food *et al.*, 1989).
   The principal specialized sources on agricultural labour, apart from the Agricultural Census, are Ministry of Agriculture, Fisheries and Food (1967) and the Ministry of Agriculture, Fisheries and Food's annual *Agricultural Labour in England and Wales*.
2. Agricultural statistics must be treated with caution (Clark, 1982c). Statistical practice varies between the component parts of the UK, each of which has its own agricultural department: UK or Great Britain data may not always be summations of exactly comparable data. Historical comparisons are hindered because of changes in the scope of the Agricultural Census. The key changes are as follows:

   (a) Expansion of the concept of 'agricultural land' principally in the uplands and Scotland.
   (b) Expansion in the number of parcels of land admitted to the Agricultural Census as agricultural holdings (principally in the late-nineteenth and early-twentieth centuries).
   (c) Improvements in the coverage and accuracy of the Agricultural Census in its early years and in 1941.
   (d) Since 1968 progressive reductions in the number of 'significant' agricultural holdings, i.e. those of sufficient size to warrant continued inclusion in the main Agricultural Census – a process happening at different times and ways in the countries of the UK.
   (e) Progressively fuller enumeration of farmers, partners, directors, their spouses, family labour and farm managers, especially since 1972.
   (f) Incompatible measures of the number of farmers and workers in the Population Census (1921 and earlier) and the Agricultural Census (from 1923).
   (g) Incompatible data on farm income as between various 'agricultural' measures of this and data from the Inland Revenue.
   (h) European Commission definitions of farm labour supply differ from those used in the UK, especially the concept of the Agricultural Work Unit.
   (i) The categorization of part-time and seasonal/workers presents some problems – they were separately enumerated in 1955 (previously linked as non-regular workers) but only in Great Britain. Northern Irish data keep them together until 1975.

3. Tax data are liable to errors and omissions, which may be legitimate (e.g. allowable deductions) or may not be, and so they must be treated with some reserve. Accountants are trained to minimize taxable income given the fiscal regime operating at the time (Hill, 1984).
4. This study, funded by the ESRC, investigated the role of farm managers and examined their position in the transformation of British farming. It involved a survey of

the activities of the eight largest agricultural land agencies and of 72 of their managers; a questionnaire survey of 60 managers selected randomly in Lancashire and Lincolnshire; an investigation of the careers of recent graduates from agricultural colleges; and a review of the public advertisements for managerial positions.

6

NEW FORMS OF EMPLOYMENT IN RURAL AREAS: A NATIONAL PERSPECTIVE
Alan Townsend

As outlined in Chapter 1, rural Britain has passed through two decades of far-reaching change, with major effects on its population, society and economy. This chapter brings together a variety of trends against one kind of benchmark – that of employment levels in rural areas throughout Great Britain. This approach is unusual because it has hitherto been difficult to aggregate such data for all rural areas, however defined. Here the government's Census of Employment data up to September 1987 (released two years later in 1989) are specially analysed for three groupings of such areas. This enables us to ask whether recent developments have placed local economic activity on a fundamentally sounder basis than previously for these areas. Indeed, has the urban–rural shift of industry and population mentioned in Chapter 1 left the economy of rural areas any different from the country at large? Have rural areas gained from the national growth of more flexible jobs for women? Do there remain, however, groups who are economically deprived, remembering the very contradictory claims made about rural trends?

Historical background

The introduction of new forms of employment was one of the the the first aims of the Development Commission when established in 1909. (This is now the Rural Development Commission, covering only England, whereas the original body covered Great Britain.) The main threats perceived, still of underlying importance, related to the social effects of depopulation and the reduced economic opportunities that resulted from falling demand for agricultural labour. At the beginning of this century, in the Census of Population of 1901, agriculture, forestry and fishing still represented 36.5 per cent of male employment in the region of East Anglia, 21.6 per cent in the South West and 16.5 per cent in the East Midlands (Lee, 1979). By 1971, however, even East Anglia had only 10.9 per cent of its working population in these activities.

A familiar picture of agricultural depression and depopulation is associated with the inter-war years of British agriculture. However, while the first three quarters of this century saw a reduction of farmers and agricultural workers from 1.33 million in 1901 to 618,000 in 1974, the greater part of the reduction of workers occurred after 1947 (Marks, 1989). The principal cause of the reduction lay in the increased use of machinery both on the land and in buildings, although some work was transferred off the farm to food and other factories. Many farmers came to do without any full-time employees, relying only on family and casual labour. However, after 1970 the rate of substitution of labour slowed down, despite the many pressures on farm incomes. These trends and their underpinnings have been described in more detail by Clark in Chapter 5.

An inheritance of deprivation

The agricultural history of the most rural counties is evident to this day in their low levels of individual weekly earnings (Champion and Townsend, 1990a, Table 11.4). The point applies to women as well as men: for instance, average adult rates per week for full-time female employees in Cornwall in 1988 stood at only £137.60, well below the national average of £164.20. The distribution of these low incomes led a number of commentators (for example, Shaw, 1979) to suggest that conditions for large parts of the rural population could be compared with the country's 'inner cities'. Housing for the poor was either decrepit or scarce, comprising limited council housing, unfurnished tied dwellings or dwellings without inside toilets. In a special survey of five rural areas, McLaughlin (1986) estimated that 20 per cent of the rural population was living in poverty.

The source of many of these problems was seen in the limited *choice* of jobs in rural areas. This 'opportunity deprivation' was compounded by 'mobility deprivation', leading particularly to the problems of finding work faced by teenagers and women who lacked their own access to a car. The proportion of women at work in rural areas has traditionally been low. The position of low-income groups can be contrasted with both the wealthy farmers (now employing few men) and with the incoming population of second-home owners and retired migrants. The latter groups may spend little in local shops and conspire against new building that would spoil their views and reduce their property values. In short, it has been possible to think of two populations in the countryside, with the wealthier ones tending to constrain – rather than enhance – the opportunities of the poorer. Local employment growth did not have to accompany the widespread recovery of population levels when it finally occurred.

A brighter picture?

It certainly seems that this picture inaccurately represents the recent growth of many rural areas. It is true that many of the efforts of the Development Commission, the Highlands and Islands Development Board, Mid Wales

Development and regional policy in developing and filling factories in the 1970s were neutralized by the national recession of 1979–82. In the end, however, rural areas fared better than industrial Britain, emerging from economic difficulty more quickly as investment resumed, and showing a level of employment growth commensurate with that of population. Far from being a feature of rural areas, *recorded* unemployment in most of the Rural Development Areas (all in England) lay below the average English rate, according to estimates made in 1987 by CACI Market Analysis for the Rural Development Commission, while, by 1990, there was virtually full employment in rural southern England.

Johnstone *et al.* (1990, p. 3) argued that the rural economy was continuing to change 'at a faster pace than before'. New 'actors' and organizations were beginning to experiment with initiatives, such as craft homes, managed workplaces, innovation centres and farm tourism. There were also a few examples of the use of advanced telecommunications to assist professional and technical developments in rural areas, amid speculation over the role of 'tele-cottages' (Dobbs, 1989). Yet if the promotional literature of development bodies is unreliable as a guide to real trends, it is worth noting that the rural areas were in fact well structured to benefit from certain national trends. Not only were these areas, for environmental reasons, attractive destinations for urban migrants and investment but their bias towards self-employment and scope for greater employment of women also gave them a potential advantage in relation to the general trends of the 1980s.

Sectoral trends in the 1980s

The central tenet of this chapter is that it is wrong to expect the replacement of farming jobs by other forms of male manual work, whether on or off the farm. It can be argued that BBC Radio's *The Archers* has got it right if that family's friends and associates are engaged in running pubs, country clubs, wine bars or antique shops, or engaged in ice-cream rounds or home dress-making. This list anticipates much better the data to follow than the forecasts of an agricultural economist. One feature of the list is its reflection of the national pattern of new sources of employment, often in low-paid work.

The country at large is experiencing marked shifts in the nature of employment from manual to non-manual work, from male to female, from full time to part time, from primary and secondary (manufacturing) to services, from permanent to temporary, and from the status of employees to self-employed (Champion and Townsend, 1990a, Chapter 6). Rural areas have traditionally shown some bias towards male employment and self-employment, particularly in agriculture, fishing and retailing. The existing literature has yet to ask whether rural areas, as one of the main loci of *increasing* national employment, are changing in line with national trends.

The answers found here are confined to the precise evidence of the Census of Employment, although this excludes the self-employed. In the following sections the issue is focused on three successively narrower definitions of rural areas. It is possible to take account additionally of farm owners, family workers, partners and directors at the level of counties of England and Wales. In a principal analysis, 22 of the 68 *counties* (including the regions and island areas of Scotland) are distinguished for their more rural character on the basis of their overall density of population (160 or fewer per km^2 in 1988). They are identified in Table 6.1, which is designed to give a preliminary view of the main growth sectors of employment – which clearly lie in fields other than traditional male manual work. A narrower definition in terms of local-authority *districts* is provided by adopting the 'remoter, mainly rural' areas of a classification developed by the Office of Population Censuses and Surveys (OPCS, 1988a, pp. 111–12), though this, too, includes some larger towns. Third, an even more focused approach is provided by use of the Rural Development Areas (RDAs) defined at *parish* level by the Rural Development Commission in 1984 but calculated here from the nearest fit of ward areas. (For the last two levels, however, data from the Census of Employment are available only from 1981 and 1984 respectively.)

Table 6.1   Percentage change in employees in individual rural counties, 1981–7

Counties (and regions of Scotland)	Persons per km^2 1988	Total %	Female %	Part time %	All services %	Tourism-related %
Highland	8	−3.1	+9.8	+20.1	+9.1	+16.0
Western Isles	11	+8.7	+16.2	+20.4	+20.3	+0.6
Shetland	16	−16.0	−1.2	+7.4	−8.9	+23.4
Orkney	20	−3.3	+11.9	+3.5	+3.9	+41.6
Borders	22	+0.4	17.7	+14.5	+11.5	+30.7
Dumfries & Galloway	23	−6.6	+6.4	+5.7	−0.2	+3.0
Powys	23	+7.2	+11.2	+8.6	15.0	+5.0
Tayside	52	−4.9	−2.4	+8.0	+2.2	−0.8
Grampian	58	+4.8	+13.7	+23.6	+13.1	+4.6
Dyfed	60	−1.9	+11.8	+12.1	+6.7	+22.7
Northumberland	60	−3.0	+12.8	+26.9	+13.9	+9.6
Gwynedd	62	−0.9	+7.3	+19.0	+4.9	+11.7
Cumbria	72	+3.4	+4.6	+5.4	+4.7	+13.6
North Yorkshire	86	+4.2	+12.6	+23.3	+11.7	+25.4
Lincolnshire	98	+2.5	+9.0	+10.0	+9.0	+4.3
Central	103	−11.6	−5.7	+15.1	−2.1	+10.7
Shropshire	115	+7.4	+13.7	+12.8	+9.2	+15.4
Cornwall	129	+6.2	+21.0	+36.3	+16.8	+8.7
Somerset	133	+5.2	+13.0	+16.5	+11.7	+15.2
Norfolk	139	+4.9	+14.9	+21.0	+14.6	+13.4
Devon	152	+4.7	+11.9	+21.3	+11.2	−2.9
Wiltshire	160	+6.5	+12.9	+3.1	+12.3	+32.3
22 rural counties	63	+2.4	+10.4	+16.7	+9.7	+11.0
Great Britain	243	−0.2	+8.1	+12.9	+9.4	+11.0

Note For composition of 'tourism-related', see Table 6.2.

(*Source*: Census of Employment (NOMIS).)

The primary sector

It might seem surprising to look for new forms of employment in the farming sector itself, but this has in fact shown a variety of kinds of employment at different places and times in the past. Today it is sometimes hypothesized that the reduction in hired full-time male labour is being compensated for by the growth of seasonal, temporary work under the same cost pressures that have been stimulating the growth of a 'peripheral' labour force in the economy at large. The main evidence for recent changes in agricultural employment has already been reviewed by Clark in Chapter 5, so here it is not necessary to do more than underline some of the main points and thereby provide the basis for comparisons with labour developments in other industries.

The single most important aspect of change in agriculture, as mentioned earlier, is the decline in the overall amount of labour input on the farm. In the ten years to 1985, the total number of people engaged in agriculture fell by 2.8 per cent, while at the same time the number of 'annual work units' (AWUs) fell by 13.2 per cent (Marks, 1989). An AWU represents the agricultural work done by one full-time worker in one year, including part-time and seasonal work as fractions of a unit.

There has clearly been a *relative* shift in the kinds of farm employment. Returns of the Ministry of Agriculture, Fisheries and Food (1989) for England and Wales, 1981–8, show a reduction of 27.4 per cent among male, whole-time hired labourers. Seasonal and casual workers (hired or family) are recorded as showing a small reduction, although Ball (1987) argues that the true extent of such 'intermittent labour forms' is heavily undercounted and that its real growth has eroded the viability of the rural economy. Regular part-time employees were stable in number from 1981 to 1988. The absolute increases occurred among part-time farmers and patterns (+13.1 per cent over the seven-year period), their spouses (+4.8 per cent) and regular part-time family workers (+14.0 per cent).

There are various possible interpretations of these data. The greatest reductions in the overall farm labour force, 1981–8, took place in the areas of the greatest overall economic prosperity, namely East Anglia (12.5 per cent), the South East (9.1) and the East Midlands (7.8), and are no doubt partly due to the availability of opportunities in other sectors. The part-time female employee remains important in the horticultural and crop-growing areas of eastern England. However, the general pattern is that farm families have been forced into greater dependence both on their own labour and in working part time in other sectors. This reflects the traditional pattern of the poorer areas of Wales and pastoral areas of the South West and Pennines with regard to the farmwork of farmers' wives. Further details of this tendency are provided by Gordon Clark in Chapter 5.

The recorded role of women in agriculture has increased in relative terms. The Farm Business Survey 1986–7 indicated that spouses did some work on 27 per cent of farms, especially on smaller farms (Gasson, 1989). It is argued that the work of farmers' wives has been ignored for too long, and that there is a need to understand the way in which farm families interpret the unequal role of women (Whatmore, 1988a). They have been involved in some of the efforts to diversity on-farm employment, such as in providing bed-and-breakfast facilities. However, the scale of farm tourism as a contribution to a farm's overall receipts is very easily exaggerated.

Off-farm employment is seen by many bodies as a more potent source of work than the government's Farm Diversification Scheme. In the USA it is estimated that over one third of farm wives and over half of all farm husbands are employed off the farm (Acock and Deseran, 1986). According to Clark in Chapter 5 of this book, 29 per cent of British farmers' wives have another job, the third highest proportion in the European Community.

Other parts of the primary sector seem to offer little by the way of extra jobs. In the most rural areas as represented by the RDAs, primary-sector employment other than agriculture fell by 2.5 per cent in the three years up to the 1987 Census of Employment. This drop, involving a loss of 500 jobs, was attributable only in small part to forestry and fishing, which together accounted for a loss of 100 jobs. The balance turns very much on developments in the coal-mining, electricity, gas and water industries and varies considerably between localities according to the decisions they make in relation to individual establishments. For instance, this category of jobs increased by over 1,000 in one RDA because of trends in open-cast mining and fuel production, while in another it fell by 800 because of trends in similar industries. These elements are sensitive to attitudes towards existing plants by industries undergoing privatization, as well as to planning decisions made partly in the interests of environmental considerations. Public policy thus has a considerable impact on these male jobs in the more rural areas.

Taking the primary sector as a whole, including agriculture, the picture presented by the 1980s is one of substantial contraction. By 1987 the 22 rural counties listed in Table 6.1 had lost 1 in 8 of the jobs they had possessed in 1981, with the pace of decline quite evenly matched between agriculture, forestry and fishing (−13.8 per cent) and energy and water supply (−11.1 per cent). The rate of loss was even higher in the 'remoter, mainly rural' districts, amounting to a loss of 16 per cent over the six-year period − the higher rate being caused by a 20 per cent reduction of employment in energy and water industries. The RDAs, by contrast, have been losing primary-sector jobs at a slower rate than our other two definitions of rural areas, with a loss of 4.2 per cent over the three-year period 1984−7, but direct comparisons are dangerous since these figures relate only to England. From this account, and bearing in mind the shift of farming jobs out of full-time work, it is very clear that it is not in the primary sector that we should look for the most striking recent openings.

Manufacturing sector

Factory employment is a new feature of many rural areas. Except for the food-processing industries and agricultural-machinery engineering of the traditional market town, an exceptional feature of Britain was the confinement of industry to the coalfields, ports and railway towns. It is true that the production of farm machinery led on to diversification into other types of engineering in many towns of midland England (such as Lincoln and Worcester). What surprised industrial geographers, however, was the arrival and growth of industry in more rural areas from the 1960s onwards. Factors in this trend included the surplus of cheap labour, the relative absence of union problems, environmental attractions and generally improved conditions of transport.

The rural areas were seen from the work of Fothergill and Gudgin (1979) as part of a systematic 'urban–rural shift' of industry and population, which involved repulsion from city and industrial areas and the greatest proportionate increase of employment in the rural areas. These far exceeded what could be attributed to the effects of 'product mix' of the inherited economic base of these areas in relation to sectoral trends. Moreover, they produced some of the greatest net job gains in far-flung areas such as Barnstaple (Devon) and Inverness (Scottish Highlands). The position in the 1987 Census, remarkably, was that manufacturing accounted for 21.8 per cent of employees in employment in the rural counties, 22.2 per cent in 'remoter, mainly rural' areas and 22.9 per cent in the RDAs of England. As the corresponding figures for Great Britain as a whole had fallen to 24.0 per cent, we can say that rural areas are barely less industrialized than the rest of post-industrial Britain!

What has quietly happened is a convergence of employment structures between urban and rural areas. This has occurred only partly because of the growth of new manufacturing enterprises or their relocation from more urbanized areas. It is principally because rural areas retained a greater proportion of their factory jobs than the other areas through the major changes of the last 20 years. The rural counties did relatively well in losing only 7.3 per cent of their factory jobs between 1971 and 1981 (though suffering job losses in food and drink and a reduction in female employment). From 1981–1987, factory employment fell by only 6.5 per cent in the rural counties and only 2.8 per cent in the 'remoter, mainly rural' areas, well below that national rate of 15.7 per cent. One element in this relatively good performance is that a number of these places were designated as 'assisted areas' over the period, either as Development Areas under the Department of Trade and Industry's 'regional policy' (Townsend, 1987) or as Special Investment Areas (from 1984, Rural Development Areas) by the Rural Development Commission. The increased programme of building small, advanced factories by the commission has been described by Chisholm (1985), and the Census of Employment data for 1984–7 enable us to measure the growth of factory employment in these particular areas.

What emerges is that while the 'remoter, mainly rural' areas then showed a recovery of 1.4 per cent, or 6,000 factory jobs, the RDAs showed an improvement of 3.8 per cent or 5,600 jobs. The RDAs where increases occurred were those in Devon, Lincolnshire, West Yorkshire, Lancashire, Cleveland and Northumberland. It would be wrong to attribute the whole of this increase to RDA status or indeed expect a precise continuation of these trends. It is sufficient to note that the rural areas are doing relatively well to hold on to their gains of factory jobs from the last few decades. There is doubt as to whether they are a seed-bed of small-firm growth: Fothergill *et al.* (1985) argue that, while this stable performance is a good achievement, the factory sector cannot provide many more jobs to meet the further problems of rural areas, even though policy–makers would welcome clean hi-tech jobs.

Tourism

Tourism-related industries have provided a major source of employment growth for rural areas in recent years (Table 6.2). Interestingly, this sector contributed almost as many extra jobs in the RDAs between 1984 and 1987 as did manufacturing, with a recorded increase of employees by 5,200, or 8.6 per cent. The hotel trade and restaurants showed a nominal reduction of employees in the RDAs (Table 6.2), although the inclusion of self-employed proprietors might show a net increase. The main increase was in 'other tourist and short-stay accommodation', the group comprising camping and caravan sites, holiday camps and miscellaneous accommodation, and this was followed by increased employment in public houses, etc. This does appear to represent some diversification of the rural economic base, especially in the direction of female, part-time opportunities.

At the level of 'remoter, mainly rural' areas, which incorporate the smaller market towns, we find that employees in employment in this group increased by 6.6 per cent from 1984 to 1987, and altogether by 22,100 (17.3 per cent) between 1981 and 1987 (Table 6.2). This is altogether a better performance than in the

Table 6.2   Change in employment in tourism-related industries in 'remoter, mainly rural' districts and Rural Development Areas

Tourism-related industries	Remoter, mainly rural				RDAs	
	1981–7		1984–7		1984–7	
	000s	%	000s	%	000s	%
Restaurants, cafes, etc.	+3.5	+17.2	+1.7	+7.9	−0.4	−3.5
Public houses and bars	+10.1	+61.4	+3.7	+16.1	+1.6	+17.0
Night clubs and licensed clubs	+1.5	+15.2	+0.2	+1.6	+0.7	+15.9
Hotel trade	+0.0	+0.1	−1.0	−2.2	−0.1	−0.8
Other tourist and short-stay accommodation	−0.1	−1.2	+2.2	+21.8	+2.5	+35.3
Libraries, museums, art galleries, etc.	+1.4	+26.6	+0.8	+12.5	+0.2	+7.7
Sport and other recreational services	+5.8	+32.5	+1.8	+8.4	+0.7	+8.8
Total	+22.1	+17.3	+9.3	+6.6	+5.2	+8.6

Note Tourism-related industries comprise Activity Groups 661, 662, 663, 665, 667, 977 and 979 of the Standard Industrial Classification 1980, as presented monthly in the *Employment Gazette*, Table 8.1. Data relate to employees in employment in September of the relevant years. RDAs refer to the Rural Development Areas of England; data for these are available only from 1984 (see text) and refer to the nearest fit of wards.

(*Source*: Census of Employment (NOMIS).)

country as a whole and is part of a clear statistical shift from traditional resort areas, such as those of Devon (Table 6.1) to less built-up coasts and upland areas, connected with a change from traditional holidays to 'day tourism' and 'short-break' weekends. As we take in successively wider areas, there is greater danger of the 'tourism-related industries' incorporating a larger proportion of local spending by residents and the incoming migrant population. The increase of 11.0 per cent in the number of employees in the rural counties in this sector is a lower rate than in 'remoter, mainly rural' areas, but this may be misleading. Nevertheless, it would appear that the continued growth of disposable income for leisure purposes will continue to generate a variety of 'tourist jobs' in rural areas, albeit part time and low-paid.

Services in general

The growth of work in hotels and catering is only one part of the transformation of Britain into a 'post-industrial' service economy; rural areas are only slightly less dependent on employment in services than is the country as a whole. What is more, many rural areas have been sharing fully in the dynamics of service-employment growth of the 1980s. What appears to have happened is that the loss of many kinds of services from villages has been overtaken by the needs of the growing overall populations in the catchment areas of most market towns, where public services have been expanded and visitors from the countryside can use a number of different private services on any one visit. The critic might argue that these jobs are beyond the reach of, for instance, village school-leavers or farmers' wives using public transport; this is, indeed, one of the conundrums of the present situation.

The distinction between rural villages and the wider area is evident in a contrast between RDAs (which are groups of parishes) and the 'remoter, mainly rural' local-authority districts. It is possible to refer only to the years 1984–7 for RDAs, but it is clear from an increase of only 3.9 per cent in these years that they did perform poorly in the service sector as a whole. The loss of jobs was led by wholesale distribution, the hospital sector (reminding us of closure programmes for cottage hospitals) and the defence sector; the net loss of retail employee jobs was quite modest and the main gains were in leisure and tourism (see above) and in community and welfare services.

By contrast, the 'remoter, mainly rural' districts showed service-sector employment increases – 6.3 per cent in 1984–7, and 13.7 per cent for the longer period 1981–7. Data show that these changes were spread across most parts of the service sector, except for transport. In terms of absolute numbers of jobs, the increases were greatest in the public sector, including welfare and community services, education and the Health Service. In terms of percentage change, however, the dominating feature is the 'banking, finance' group, led by the sub-sector of 'business services', which in southern rural areas (such as parts of Norfolk and Devon) grew at a faster rate than the national average, as did retailing and wholesaling.

Overall employment trends

In whatever way we define rural areas, when we take the picture of all sectors together, we find strong employment growth, which is generally not attributable to these areas' mix of growing and declining sectors (see 'shift-share analysis' in Champion and Townsend, 1990a, Table 11.5). From 1971 to 1981 and from 1981 to 1987, job losses in the rural counties in the primary, manufacturing and construction sectors were totally offset by a large increase in service employment (Table 6.3). There was some diversification into 'other manufacturing' in the 'remoter, mainly rural' areas (Table 6.3), which contributed to a net increase of no less than 5.0 per cent in all employees in employment there between 1981 and 1987, compared with a small national decline. In the deep rural areas represented by the RDAs, the growth of manufacturing and tourism in 1984–7 was somewhat stronger than for the 'remoter, mainly rural' areas over the same period, but more sluggish trends in a number of services reduced overall growth to 3.1 per cent (Table 6.3, last column).

At this point, however, we must enter the caveat that, as in agriculture, in terms of total 'work hours' the trend for total employment is less favourable than is suggested by changes in the size of the active labour force. Moreover, there is a stronger swing from male to female employment than in farming. The evidence in Table 6.3 (bottom panel) clearly indicates that the number of male and full-time employees remained relatively static in rural Britain during the 1980s, though this itself is a remarkable achievement in the context of national decline in the numbers in these two groups.

The striking feature coming out of this analysis, then, is that the rural areas turn out to be a leading part of Britain for the national swing towards the female and part-time forms of employment, albeit starting from lower levels than the rest of the country (Townsend, 1986). In the 'remoter, mainly rural' areas, part-time employment increased by 20.1 per cent in 1981–7 and 10.2 per cent in 1984–7. There is little doubt about the general consistency of these data, which are provided by employers who operate PAYE arrangements; there is probably some under-recording, but no reason to suppose that this markedly reduced during the 1980s.

At the same time, however, it must be stressed that, as far as female employment levels are concerned, this pattern of change leaves the rural areas in much the same position relative to the rest of the country as before. In the 1981 Census of Population it was precisely these rural areas that showed the lowest ratio of economically active to adult (16 and over) females. Percentages of 34.3 per cent (Western Isles), 34.8 per cent (Cornwall) and 35.8 per cent (Gwynedd) compare with an overall total of 38.1 per cent labour-force participation by females in the RDAs, contrasting markedly with the Great Britain average of 45.5 per cent. No later statistics of activity rates are available at this scale. It would be safe to assume, however, that even in the RDAs the proportion of females in some kind of work will now exceed 40 per cent – still below average but improving.

People in the Countryside

Table 6.3 Employment structure and change for three definitions of rural areas

Employment groupings	Structure 1987 (% total)				Change 1981–7 (%)			Change 1984–7 (%)	
	Great Britain	Rural counties	Remoter, mainly rural	RDAs	Great Britain	Rural counties	Remoter, mainly rural	Remoter, mainly rural	RDAs
Agriculture, forestry and fishing	1.5	4.7	7.7	11.2	−11.4	−13.8	−13.5	−7.4	−4.7
Energy and water supply	2.3	3.2	1.7	2.7	−28.2	−11.1	−20.3	−7.4	−2.2
Manufacturing, incl.	24.0	21.8	22.2	22.9	−15.7	−6.5	−2.8	+1.4	+3.8
Minerals, metal-processing	3.2	2.6	2.8	4.8	−24.2	−14.6	−11.2	−13.3	−7.2
Metal goods, engineering and vehicles	10.9	8.5	7.9	7.1	−18.7	−9.2	−5.2	+2.0	+4.2
Other manufacturing	9.8	10.7	11.5	10.9	−8.5	−2.1	+1.3	+5.5	+9.1
Construction	4.7	5.4	5.4	5.7	−8.1	−11.9	−9.7	−0.5	+11.9
Services, incl.	67.4	64.9	63.0	57.5	+9.4	+9.7	+13.7	+6.3	+3.9
Distribution, hotels, catering, repairs	20.0	22.2	22.5	22.2	+3.7	+3.9	+8.2	+1.7	+0.8
Transport and communication	6.0	5.4	5.4	4.5	−8.4	+1.1	+5.0	+6.6	−0.8
Banking, finance, etc.	10.9	7.1	6.3	4.9	+33.4	+27.4	+32.1	+8.9	+6.7
Other services	30.5	30.2	28.9	25.9	+10.7	+12.4	+9.4	+16.7	+7.0
Total, of which	100.0	100.0	100.0	100.0	−0.2	+2.4	+5.0	+3.4	+3.1
Males	54.3	54.6	55.0	56.8	−5.4	−3.4	−1.7	−0.3	+0.7
Females	45.7	45.4	45.0	43.2	+8.1	+10.4	+14.4	+8.3	+6.3
Full time	76.1	74.3	73.6	73.3	−3.1	−1.8	+0.5	+1.1	+1.2
Part time	23.9	25.7	26.4	26.7	+12.9	+16.7	+20.1	+10.2	+8.4
Tourism-related	6.0	7.8	7.9	9.6	+11.0	+11.0	+17.3	+6.6	+8.6
Total (000s)	21,271.0	2,750.4	1,896.8	672.2	−38.0	+64.7	+90.1	+61.9	+19.9

Notes See text for definitions of rural areas and Table 6.1 for list of rural counties; see also notes in Table 6.2.

(*Source:* Census of Employment (NOMIS).)

Conclusion

The general pattern to emerge from this chapter, therefore, is that a relatively depressed view is justified for male employment. The general reduction of manual work on the land and elsewhere has, in general, been fully offset – but no more than this. The brighter picture applies chiefly to females, whose relative importance has increased, so that in 1987 they represented a total of 45.4 per cent of employees at work in both the rural counties and the 'remoter, mainly rural' areas and 43.2 per cent in RDAs. This is a striking development: it does contribute to real economic change in terms of business growth and improved services, but must be measured against the recent pattern of women remaining at home and in terms of low incomes and short hours.

Acknowledgements

Thanks are due to Dr Janet Townsend for suggestions on the role of gender, especially in agriculture; and to the staff of the National Online Manpower Information System (NOMIS) for their programming.

7

WOMEN IN THE RURAL LABOUR MARKET: A POLICY EVALUATION
Jo Little

Much publicity has been given of late to the structural changes taking place in the economy and society of contemporary rural Britain. Some of these changes have been taken up by other chapters in this volume. In response to a continuing reduction in the number of people employed in agriculture and related industries, attempts are being made to broaden the economic base of rural areas. Policy makers, the business community, developers and rural people are beginning to recognize the need for diversification in order to expand the number of job opportunities available to indigenous people and to maintain diverse village communities. Despite the high political and academic profile of these changes, however, we really know very little about their relevance for and impact on different groups living in the countryside – for example, for the young, the long-term unemployed, the unskilled. We know even less about the impact of current policy on the fortunes of these different groups.

The purpose of this chapter is to look specifically at the implications of employment-related policy for the lives of one group in the rural community, women. Women have been identified as particularly 'in need' in relation to rural employment (see Development Commission, 1987). But, again, little is really understood of the extent and nature of such need or of the relationship between need and policy. This chapter questions the effectiveness of contemporary policy initiatives in either recognizing or addressing the constraints experienced by women who are, or would like to be, involved in the rural labour market. It looks in particular at the degree to which job-creation policies have targeted women as a special group. The chapter draws on material from both a national and a local level and includes preliminary data from a broad-based study of women's employment in rural areas.[1]

In seeking to examine issues of women's employment in rural communities the lack of existing information is immediately apparent. Rural researchers have devoted very little time in relation to their urban counterparts to the identification and documentation of women's roles and experiences generally. More importantly has been an associated failure by those studying the rural economy and society to recognize the contribution of gender to the evolution and maintenance of social relations (exceptions include many of those working on agricultural transformation). There is, therefore, a need for the generation of a body of detailed and original information on women's lives in rural areas. Such information should not, however, be collected in isolation but should be used in (and informed by) a much broader analysis of gender relations as they affect the lives of those living in rural communities.

This is not primarily a theoretical chapter and it is not the intention here to become immersed in a critique of various theoretical debates surrounding the interpretation of gender relations or in their relative value in helping us to understand social relations and women's roles (a huge range of material is available on these issues – see Little, 1987a, for discussion). It is important, however, to stress that the information discussed here has been collected and analysed within a framework that recognizes and (it is hoped) informs these debates. Central to the arguments raised here is that women's employment experiences – their access to paid work and the conditions they encounter within the labour market – are a function of the operation of patriarchal gender relations. Policy, it is asserted, not only reflects gender relations but also constitutes a key mechanism for sustaining and reinforcing existing inequalities based on gender. The analysis of policy thus cannot be divorced from either the derivation or the consequences of gender relations.

Constraints on women's employment

Before embarking on a detailed analysis of employment policy and its effects on women's job opportunities, it is helpful to look briefly at the nature of women's current involvement in waged work and at the major constraints surrounding their participation in the rural labour market As has already been mentioned, there is a serious lack of existing data on these issues. The few studies that have been undertaken have demonstrated women's employment in rural areas to be characterized by poor wages, low levels of skill and a lack of choice. Activity rates among women are traditionally lower in rural than they are in urban localities, particularly in the most remote areas, and work performed by women is often insecure (casual or seasonal) and non-unionized.

In his work on rural deprivation, McLaughlin (1986) identified a high level of part-time employment amongst women. His data, drawn from a sample of five case-study areas, revealed that 32 per cent of employed women living in rural areas were engaged in part-time jobs, as opposed to 21 per cent nationally. A more recent study of women's employment in rural Hampshire (Collins and Little, 1989) demonstrates the even greater involvement of women in part-time employment – a total of 74 per cent of employed women were found to work in jobs that were classified as part time. This has several implications. In the first place, women working part time naturally have lower incomes than those in full-time jobs. In the Hampshire study, for example, it was found that women who were employed part time earned an average of £3,040 per annum, with half earning less than £2,400. Women employed full time, however, earned an average of £9,874, with the highest salary over £22,000 a year. Obviously these figures are not directly comparable since not all the women employed part time were working the same number of hours. They do, however, provide some indication of the poor levels of pay received by part-timers.

Second, it is widely accepted that part-time employment is generally characterized by poorer working conditions than full-time employment (Robinson, 1988). People employed for less than 17 hours per week are not subject to the same conditions surrounding holiday pay, sickness benefits, pension and other workers' rights as those in full-time jobs (Beechey and Perkins, 1987; Little, 1990). As the majority of part-time employment is undertaken by women, as a group women inevitably are more affected by these poor conditions. The problems are perhaps most serious in rural areas, where the lack of choice means that women are more dependent on the existing opportunities and where the very low rates of unionization among women make them powerless to object. Again, the Hampshire study provided evidence of the poorer conditions experienced by part-time workers: 56 per cent of women employed part time did not receive sick pay and 58 per cent had no paid holiday entitlement.

Comparing different localities, McLaughlin (1986) also discovered women in full-time employment in rural areas to be generally more poorly paid than their urban equivalents. Conditions were found to be particularly bad in the manual sector with over 77 per cent of women employees in rural areas being defined as low paid (as opposed to about 66 per cent nationally). Closer inspection of existing (albeit rather limited) data reveals that women's wages, like those of rural workers generally, are polarized between the very high and the very low. Unlike men, however, employed women are very markedly concentrated in the lower-wage brackets.

Explanations of rural women's employment experiences have tended to rest largely on structural criteria. Their concentration within certain sectors is seen simply as a function of the rural labour market and of the limitations imposed by lack of job provision. Clearly, women have been constrained by the low levels of job availability and it is one of the first requirements of policy that the *number* of job opportunities in rural areas be increased. But women's involvement in paid work is not simply a case of supply. Their access to jobs is conditioned by much broader characteristics of their roles – especially within the domestic sphere. As primarily (and often exclusively) responsible for the running of the household and the daily care of its members, women are often constrained in terms of the number of hours they can work outside the home. Moreover, their position as secondary (as opposed to primary) income earners means that they generally occupy second place in relation to household transport and are thus often dependent on public transport. In addition to these practical constraints, women experience a range of ideological expectations in connection with their role in the rural community and domestic household (Little, 1987b), which impose further restrictions on their access to paid work.

It is in this context, with a recognition of the broader constraints affecting women's access to, and experiences in, employment, that policy must be formulated and evaluated. As noted, if policy is to address the true problems faced by women in the rural labour market it must do more than simply look to the creation of opportunities but must recognize the need for particular types of opportunities. Even if diversification does result in new jobs in rural areas, these jobs may not be accessible to women. Many women are working in jobs that are neither what they want nor what they are trained to do. It is the problems of *these* women that must be addressed as well as the needs of those wishing to enter the labour market. In the analysis of policy that now follows, these issues are explored and emphasis is placed on the success of policy in identifying and addressing the specific constraints operating on women.

Evaluation of national and local policies

National policy

Before looking at the formulation and implementation of employment policy at a local level, it is useful to consider briefly the national context and to evaluate the extent to which women's needs have been addressed within a national-policy agenda. Specific initiatives at this level are conspicuous by their absence and, while national agencies such as the Rural Development Commission have acknowledged the high levels of female unemployment in some rural areas, their acknowledgements have not been accompanied by policies or concrete suggestions for change. The poor conditions under which many women are employed in rural communities are similarly recognized but not specifically addressed. The assumption would appear to be, at the national scale of policy making, that the particular needs of women in the rural labour market do not require targeting through separate initiatives but that women will benefit, in the same way as men, through the general diversification of the rural economy.

In addition to failing to provide specific employment initiatives, the State, it may be argued, is responsible for actually *maintaining* the conditions under which women are employed and reinforcing their domestic role. For example, through the organization and maintenance of the social-security system, the State has increased women's dependence upon their husbands, restricting their direct eligibility for social-security benefits and making them vulnerable to use as cheap labour power. A range of other State provisions aimed at alleviating the poverty of working-class households (for example, child benefit, family income supplement, temporary sickness benefit, etc.) also act to 'bolster a specific family household system in which the wife is assumed to be dependent and in which she is assumed to provide household services for all the other members' (McIntosh, 1978, p. 271). The State's role, while not exclusive to rural areas, is nevertheless an important influence on women's participation in the rural labour market. In addition, the State is directly and consciously responsible for reinforcing women's domestic role in the neglect of public welfare provision and a reliance on 'community care'. This issue does have more relevance to rural areas where the low base level of, for example, hospital facilities necessitates a greater dependence on such community provision (i.e. unpaid labour by women).

A related area of State influence over employment participation in rural areas concerns training. Women, it is argued, are disadvantaged in terms of job opportunities because of a lack of training. This is particularly true of married women wishing to return to waged work after the birth of a child, but it can also apply to young women. In a study of Youth Opportunity Programmes in rural areas, for example, McDermott and Dench (1983) found that boys were more likely than girls to gain some form of work experience or training while at school and that, while 75 per cent of boys went into skilled work on leaving school, only 33 per cent of girls did so. The lack of training available to women living in rural areas ensures that they are restricted to occupations of low skill and poor pay. The training that does exist, moreover, frequently serves to reinforce women's 'traditional' domestic skills. According to McDermott and Dench (*ibid.*), Youth Opportunity Programmes in rural areas offered boys training across all sectors, but gave girls mainly clerical and caring placements almost exclusively within the service sector.

Local policy

The relationship between policy and employment is perhaps more apparent at the local level of analysis. The Rural Development Commission has, in the formulation of Rural Development Areas (RDAs), sought to stimulate employment opportunities in rural localities and generate growth in areas demonstrating decline. While this has been a national initiative (covering 28 areas across England), its implementation, and, indeed, the specific details of the policy, have been worked out locally (for more details of RDAs as a policy mechanism, see Chapter 11, Bowler and Lewis). As this is one of the few, broader, employment-related initiatives to be aimed at rural communities, it is important to consider the extent to which the particular needs of women in the labour market have been addressed by the various Rural Development Programmes (RDPs).

The RDAs themselves demonstrate high instances of the kind of traditional characteristics associated with women's employment described earlier. In particular, the range of employment opportunities available to women within the RDAs has been, and continues to be, severely restricted. A serious lack of jobs in the manufacturing sector has meant a disproportionate dependence on primary industry (generally agriculture, but also extractive in the Forest of Dean and Durham) and on the service sector. In Northumberland RDA, for example, only 14 per cent of all jobs in 1984 were in manufacturing (as opposed to 26 per cent for the country as a whole) and in North Yorkshire RDA only 19.5 per cent fell into this sector. Women's employment within the RDAs is concentrated in part-time, seasonal or casual jobs: in North Yorkshire RDA, for example, 53 per cent of all female employment fell into such categories. As argued earlier, work in these types of jobs is more likely to be of an unskilled or semi-skilled nature.

In relating the employment experiences of rural women directly to policy initiatives within the RDAs, three particular issues can be examined as to whether or not RDA strategies

1. include any recognition of female activity rates or of unemployment among women;
2. include women's employment provision as a policy objective; and
3. include concrete initiatives for the generation of employment opportunities for women.

Such an analysis has been carried out in relation to 20 of the 28 RDA strategies, with information being drawn largely from the broad strategy documents and the most recently produced 'rolling programmes' for each area.

Clearly, as has so often been stated in the analysis of policy, simply studying written documentation does not necessarily provide a complete, or even accurate, picture. In this instance, however, the analysis was intended purely to provide an indication of the degree to which RDAs recognize the scale of the problems facing women in relation to rural employment and to obtain an idea of their commitment to tackling these problems.

In looking at the RDA strategies, it soon becomes apparent that the specific employment problems experienced by women generate little direct or separate attention. Table 7.1 summarizes the performance of the different RDPs in accordance with the criteria outlined above. In relation to the first issue, 12 of the 20 programmes analysed in the table recognized the particular problems of low activity rates among women. Typical of such programmes was North Yorkshire RDP, which described the characteristics of the RDA itself as follows:

> Low density of population, under employment, long journeys to work, lack of readily available training and re-training opportunities, *limited opportunities for female employment*, lack of job variety, entry into routine jobs below individual ability, out-migration of young people and others of working age, imbalances in age structure of the population, low income levels, a steady decline in the quality and range of services and a heavy reliance on grants and subsidies.
> (North Yorkshire County Council, 1985, p. 1, emphasis added)

Like many of the other strategies, the lack of employment opportunities for women is recognized but simply incorporated into a general statement on the 'problems' of the area. Women's needs as a separate issue fails to re-appear in the majority of cases. Indeed, only three strategies follow up the recognition of women's employment problems with any related policy objective. The three RDPs are West Somerset (1986, although the explicit commitment to policy is lost in later versions of the document), Northamptonshire (1987) and Gloucestershire (1985a). Both the latter include women's needs within broad strategy statements on employment. The Gloucestershire RDP, in an opening paragraph on employment objectives, states that (*ibid.* p. 2)

> there is a pressing need to improve job opportunities in the area for both men and women and in particular:
> (a) to reduce the very high levels of unemployment

Table 7.1   The consideration of women's employment issues within Rural Development
Programmes

Area	Mentions of women's low activity rate or unemployment	Policy objective on women's employment	Women's initiatives
Cheshire	√		
West Somerset		Early plans	
Norfolk			
Derbyshire			
Romney Marsh			
Shropshire			
Cornwall	√		
Lancashire			
Durham	√		
Devon	√		
Northumberland	√		
Humberside	√		
Dorset	√		
North Yorkshire	√		
Northamptonshire	√	√	
Cumbria			√*
Suffolk	√		
Gloucestershire	√	√	
Fenland			
North-East Staffordshire	√		

Note * Initiative *not* for employment.
Tick indicates that Rural Development Programme includes the item specified in the
column heading.

    (b) to widen the range of job opportunities
    (c) to stimulate growth in the service sector which is poorly represented in the area
    (d) to provide job opportunities for women and school-leavers who make up large
         components of the unemployed.

Similarly, the objectives provided in the Northamptonshire RDP (1987, p. 1)
state a commitment

> To provide employment for a greater proportion of young people entering the labour
> force each year.
> To provide employment for those currently and potentially unemployed.
> To provide greater employment opportunities for women residents of working age.

The poor record on policy objectives aimed specifically at addressing women's
employment problems in the RDAs is not surprisingly mirrored in the context
of concrete initiatives. None of the 20 RDA annual programmes included
the development of actual initiatives aimed at the generation of employment
opportunities for women. Indeed, only one programme contained any women's
initiatives at all – this was the Cumbria RDP (1987), which included a leisure
development specifically aimed at women. The lack of employment initiatives

within the RDA can be equated with a more general neglect of women's programmes within the public sector in rural areas. Halford (1987), in a study of women's initiatives in local government, remarks on the urban bias of such initiatives, suggesting that they are almost exclusively associated with areas of Labour control, notably in towns and cities. She concludes, on the basis of a national survey, that 85 per cent of all known initiatives have taken place in authorities that are at least large towns, if not cities or metropolitan conurbations. Clearly, the existence of a women's committee at local-government level cannot be automatically associated with superior employment opportunities or conditions for women. Such initiatives do, however, generally denote a certain level of understanding about the constraints operating upon women, and a commitment to improving both access to, and experiences within, employment.

It is not only the absence of specific policy initiatives that has contributed to women's current position within the rural labour market. Existing employment policies in rural areas do not only fail to prioritize women's needs but also act, it may be argued, to reinforce traditional characteristics of women's participation in the job market. Emphasis is placed, for example, on the provision of workshops and small industrial units as most 'appropriate' for the rural environment. Women, where employed, are frequently working in isolation or with very few other employees. Consequently, the social advantages associated with waged work are lost, while working conditions may be abused due to the powerlessness of the workforce.

Sectorally, the main emphasis within rural job creation has been on 'craft industry' – light manufacturing and service-sector employment. Within these sectors women are generally employed as carers or secretarial/clerical staff or in repetitive low-skill work, such as packaging. Moreover, the current support for self-employment in rural areas is disadvantageous to women, for while some may find it liberating to organize their own employment, for a variety of reasons most have not the access to money or to some of the formal and informal networks that help in the provision of advice, training and even premises, to allow them to start their own businesses.

Research undertaken in rural Suffolk has attempted to explore these issues in greater depth, by focusing at the local level on the relationship between women's employment opportunities and the impact of both specific policy initiatives and broader-based tendencies and priorities for employment generation. The remaining section of this chapter will provide a brief description of the early results of this research. (For further details, see Little and Ross, 1990.)

The impact of employment policy: the case of the Suffolk Rural Development Area

The Suffolk Rural Development Area comprises 110 parishes within which, in accordance with criteria for RDA designation, there is a recognized need for social and economic regeneration, with the relatively high rates of unemployment and the narrow economic base of this area standing in clear contrast to more urban parts of the county. Suffolk County Council draw particular attention within the RDA programme to the imbalanced employment structure and the dependence on agriculture and service industries within a limited number of firms (Suffolk County Council, 1987). In common with a number of other RDA programmes (as shown in Table 7.1), the Suffolk programme recognizes the shortage of local jobs for women and the relatively high female unemployment rates within parts of the RDA.

Like other RDAs, however, the Suffolk RDP does not follow up the acknowledgement of high levels of female unemployment with any specific policy initiatives. There are no policies or examples of projects within the programme aimed directly at women, either in the context of employment provision or in other (potentially related) fields.

Priorities for the creation of employment opportunities within the RDA rest largely on the provision of small workshop units. Need is perceived over-whelmingly in the context of workspace availability and consequently solutions revolve around the supply of buildings and land rather than actual jobs. Between 1979 and 1985, the RDP provided finance for the creation of 19 workshops (100 per cent funding). This led to a total of 65 new jobs. Since this time other schemes, such as a partnership between the Rural Development Commission and the local authority, have resulted in further workship provision within the RDA – only, however, for very small-scale development. Provision of premises appears to be targeted largely at new businesses, with an emphasis on 'starter units'.

As mentioned above, the direction of policy and funding so disproportion-ately towards the workshop sector and the resulting emphasis that it places on self-employment can be disadvantageous to certain groups within the rural population, particularly women. Discussion with the Council for Small Industries in Rural Area (COSIRA) (now part of the RDC's 'business service') suggested that the take-up of starter units and the establishment of small business by women in their own right within the rural parts of the country was low. In March 1988, the Suffolk branch of the COSIRA had 65 businesses on their books, of which only five were run by women (in name).

Another problem relating to the current policies for employment generation in the Suffolk RDA is the location of new developments and, consequently, of potential job opportunities. These are largely concentrated within the small towns and larger villages, despite being part of a *rural* development programme. As has already been mentioned, one of the most important constraints encountered by women wishing to enter the rural labour market is that of physical access. Even apparently short journeys are difficult or impossible, given the very poor levels of public transport that exist in many rural areas and the additional constraints (such as taking children to school) which impinge on women's employment participation. This point is well illustrated in the Suffolk study area where women living in certain parishes were prepared (or possibly obliged) to work for low wages and in very poor conditions in a local chicken-processing firm and on a fruit farm because the work was accessible. The parishes themselves are not particularly remote in pure distance terms, being on the western edge of the RDA and thus relatively close to larger employment centres such as Bury St Edmunds. The greater flexibility offered by local employment, together with the availability of a special bus provided by the owners of the fruit farm, were advantages that outweighed the superior pay and conditions potentially available elsewhere.

Coupled with the issue of accessibility is that of job training. Women, it has been argued, are disadvantaged within the labour market due to a limited access to job training *per se* and, perhaps more importantly, by a rather narrow interpretation of the 'appropriate' forms of skills and training for women. While there is an indication that current conditions in the job market generally are forcing employers and policy makers to take a new look at training provision, especially for older women, women living in rural communities still experience particular difficulties in getting access to the type of training they would like. In addition, job training (whether it be within or prior to employment) is more commonly associated with full-time as opposed to part-time employment and, as noted earlier, greater proportions of women employed in the rural labour market work part time. The research undertaken in Hampshire and referred to earlier found that, while 56 per cent of women employed full time had undertaken some kind of formal training in connection with their current job, only 27 per cent of women employed part time had done so (Collins and Little, 1989).

Access to training opportunities for those living within the Suffolk RDA has been recognized as a particular problem (Suffolk County Council, 1987). It is only young people as a group who are identified as in need of special attention by the RDP, however, and the particular difficulties of women are not discussed. According to a training consultant working within the area, there is a high latent demand for job training among women, particularly those aged over 35. A series of preliminary training workshops organized within the area (and aimed at improving both employment eligibility and education) among women from a variety of backgrounds had yielded high levels of interest and attendance.

The recognition of women's need and eligibility for employment training is, on one level, encouraging – even if initiatives are slow to penetrate the less accessible rural areas. Care must be taken, however, to direct training opportunities to meet the specific needs of different groups of women and to provide new skills rather than simply reinforcing existing patterns within the labour force. The training courses established in Suffolk were clearly an important focus, bringing women with similar needs together, but were not designed to provide an ongoing programme of skill development. Like other rural initiatives these courses were given a low priority by the local education authority and consequently received little finance, publicity or staff provision, relying almost exclusively on the enthusiasm of the individual women responsible for setting them up. Unlike larger-scale, high-profile training courses for women, such rural-based initiatives cannot easily develop opportunities for women within non-traditional areas. They are limited, by and large, to addressing more established areas of 'women's work' (such as word-processing and cooking) and consequently reinforcing stereotypical ideas of what is appropriate employment for women.

Finally, on the issue of job training, it is important that those responsible for setting up training programmes understand the very basic level of need that many women, especially in rural areas, experience. Training must address not simply the question of job skills in a conventional sense, but also look more widely at the problems of women in relation to, for example, confidence and personal development. Many women returning to work after a long period away from the labour market, or perhaps entering formal employment for the first time, are considerably lacking in confidence and require 'training' in such areas as job search, application completion and general self-awareness as much as in specific job skills. It is important, therefore, not to see job training *per se* as the panacea for the problems experienced by women wishing to participate in the rural labour market, but instead to view it in the context of other constraints operating upon women. Attempts must be made to develop training in response to the particular needs and circumstances of rural women and not in an isolated vacuum.

Conclusion

This chapter has examined the relationship between employment policy and women's experiences within the rural labour market. The central question has been the extent to which the particular constraints and problems operating on women's participation in rural employment have been recognized and addressed at the various levels of policy making. The work has drawn on both established research and original information; and while there is much further work to be done on this topic, a number of interesting conclusions can be drawn.

In electing to concentrate specifically on policy the chapter has, it may be argued, overlooked the more localized impact of the implementation of particular decisions and initiatives and consequently failed to pick up on real changes on the ground that may affect women's access to and experiences within employment. It is also important, however, to take a more strategic view of the issues and to consider the broader policy framework within which specific initiatives are located. Only by looking at this level can the wider commitment to and understanding of women's employment opportunities be assessed.

This chapter's analysis of policy at both national and local levels reveals little by way of firm commitment by policy makers to addressing the problems faced by rural women in relation to employment. Some degree of recognition is afforded to the specific problems of unemployment among women within RDAs, but given that low activity rates were one of the main criteria for the establishment of RDAs, this is hardly surprising. More pertinent is the absence of policy initiatives and concrete projects that accompany the apparent awareness of the constraints operating upon women. The unwritten belief among policy makers would appear to be that the provision of employment opportunities for women will develop from a more general encouragement of economic growth *per se* in rural areas. Such a view, however, fails to recognize the particular nature of the problems faced by women and the need for specific tailor-made solutions.

In addition to drawing attention to the basic absence of policy responses, the chapter has also argued that the emphasis of existing policies of economic generation in rural areas serves to reinforce the dominant characteristics of women's employment. Opportunities for women's participation in the rural labour market need to be addressed not simply in the generation of new initiatives but in the challenging of existing priorities. Parts of this process clearly involve an attack on the more general conditions under which women enter the labour market – conditions determined by Welfare State and employment policy. Other parts, however, relate more directly to the characteristics of economic development and employment provision in *rural* areas. It has been argued here that a preoccupation with scale and character has limited the range of job opportunities available to women and indirectly encouraged the segregation of employment. Women have been forced to rely on traditional forms of employment, many of which are part time, insecure and low paid.

The key point is that successful policies cannot be formulated without a proper understanding of the constraints experienced by women in all aspects of their lives and of how these areas interact. It is not sufficient simply to provide 'work places' or even jobs *per se*. Policy must address issues of access and childcare and thereby acknowledge that women's participation in waged work is not simply a function of the operation of the labour market. Only by a commitment to change throughout women's lives will progress be made in terms of their involvement in waged work. Again, while these are very much issues around local service provision, they need to be recognized at a broad strategic level such that they may be taken on board across a spectrum of rural policy. Clearly, this discussion of existing initiatives and attitudes demonstrates that there is still a very long way to go.

Note

1. This study entitled Women and Employment in Rural Areas, was carried out in 1989–90 and was funded by the Rural Development Commission and the Royal Agricultural Society of England. It looked at the experiences of women in the rural labour market in three areas of Britain – East Wiltshire, North Cornwall and North East Derbyshire.

8

THE ELDERLY AND DISABLED IN RURAL AREAS: TRAVEL PATTERNS IN THE NORTH COTSWOLDS

Robert Gant and José Smith

In the 1980s the shift of political ideology to the right in Britain and severe restraints on public expenditure have combined with demographic trends to raise a series of important questions regarding the welfare of the elderly and disabled (Warnes and Law, 1984; Bell and Cloke, 1989). These issues have been addressed from a geographical perspective by studies of changes in the spatial distribution of the elderly, research on migration patterns and processes, assessments of spatial variations in well-being, and policy-orientated contributions on the geography of service delivery (Bailey and Layzell, 1983; Rowles, 1986; Champion *et al.*, 1987; Warnes, 1982, 1987).

This agenda will remain in focus given reliable forecasts that, in the period 1981–2001, the proportion of the British population aged 65 or above will rise from 15 to 17 per cent, the cohort aged at least 75 will increase by 42 per cent, and those aged over 85 will double (Family Policy Studies Centre, 1988). Public awareness of this situation and its financial ramifications has been raised by informed media coverage of two recent publications: first, the OPCS survey, *The Prevalence of Disability Among Adults* (1988b), which defined an adult disabled population of 6.2 million and itemized forms of disability and their implications for everyday life; and, second, the white paper, *Caring for People* (HMSO, 1989), which declared the government intention to encourage local authorities to offer care contracts for the elderly and disabled to the private and voluntary sectors (Johnson, 1988; *Guardian*, 1989). It is now recognized, moreover, that chronic illness and serious mental and physical disability are becoming compressed into a shorter, and later, stage of life. This is compounded by gender differences in life expectancy whereby the predicted transition of the 60-year-old cohort in 1985 to age 80 is set at 35 per cent for males and 57 per cent for females (Thompson, 1987; Wells and Freer, 1988).

These stark demographic trends and policy measures have serious implications for the future delivery of health and welfare services (Fearn, 1987; Haynes, 1987). Attempts have consequently been made to define and measure 'risk-groups' among the elderly on the basis of social, economic, housing and family circumstances. Characteristically, these studies have focused on the elderly in an urban environment (Ford and Taylor, 1983; Taylor, 1988). However, the problems have been equally as acute in rural Britain, where there has been a widespread decline in the provision of community services and public transport in the post-war period (McLaughlin, 1986; Association of District Councils, 1986; Lowe, Bradley and Wright, 1986). These changes in accessibility have resulted in a progressive deterioration in the welfare of the elderly, many of whom are disabled (Buchanan, 1979, 1983; Fennell, 1988). In areas such as the Cotswold District of Gloucestershire, where the social services are organized on a 'patchwork' basis, the effective planning and co-ordination of welfare services demands a detailed appraisal of travel patterns (Frye, 1986; Smith and Gant, 1982, Gant and Smith 1988). This chapter describes a study that meets this requirement and focuses on the movement patterns of the elderly and disabled in eight rural parishes in the centre of Cotswold District (Figure 8.1). These parishes were selected to capture differences in location relative to the principal roads and towns; contrasts in population size; available services and public transport provision; and to represent variations in local demographic trends.

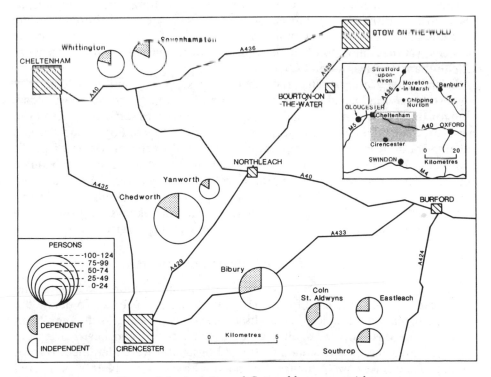

Figure 8.1   Location of Cotswold survey parishes

The regional setting

The popularity of the Cotswolds for retirement and the outmigration of younger elements have distorted the age structure of the region (Dunn, Rawson and Rogers, 1981). In 1981, for example, 27 per cent of the 27,000 residents in the North Cotswolds Social Services Area were pensioners, and 8 per cent were aged over 75. Overall, 27 per cent of the pensioners lived alone; this proportion increased to 39 per cent for those aged 75 and over. This imbalance is accentuated by marked spatial variations in the proportion of pensioners in the 63 parishes of the Cotswold District. This proportion ranges from 6 to 41 per cent, with a median parish value of 24 per cent. Figure 8.2 emphasizes this variation, and shows the scattered distribution of disabled pensioners by the type of settlement recognized in the County Structure Plan (Gloucestershire County Council, 1986a). These distributions have important welfare implications. Not surprisingly, the elderly have substantially inflated the case loads of the social-services team working from Moreton-in-Marsh, and in 1984–5 accounted for 68 per cent of the 1,400 casework referrals (Gloucestershire County Council, 1986b).

Alongside these demographic changes there has been a widespread reduction in the level of community services and their distribution in the settlement network. At present a balanced but minimal selection of health care, professional and retail services is provided only in five small market centres – Chipping Campden, Moreton-in-Marsh, Stow-on-the-Wold, Bourton-on-the-Water and Northleach. Here, too, are located general-practitioner-run hospitals, dentists, dispensing chemists, homes for the elderly, sheltered housing schemes and the main police stations. For higher-order health and shopping services, journeys have to be made to major towns outside the region, such as Cheltenham, Cirencester and Oxford.

In the survey area Bibury and Chedworth support a full complement of low-order services including a part-time doctor's surgery, grocer, post office, club for the elderly and parish hall. All the smaller villages surveyed lacked even a grocer, although mobile vendors provided milk and bread. Population size and relative location were again important with regard to the provision of public transport: Whittington, situated 1 km north of the A40, has no scheduled service; Yanworth and Sevenhampton have one weekly service; Eastleach and Southrop shopping services two to three days per week; and Bibury, Chedworth and Coln St Aldwyns have services on at least four weekdays. Not one of the survey villages had the benefit of a weekend or evening service.

Disability in a household context

Students from Kingston Polytechnic and the Community Enterprise Programme assembled the data base used for the study in the spring of 1982. To define the survey population of elderly and disabled adults a short, self-completion questionnaire was delivered to all households in the survey parishes. These were collected and checked and some additional households were identified by local inquiry. Field interviewers then visited each address to complete a pre-tested questionnaire.

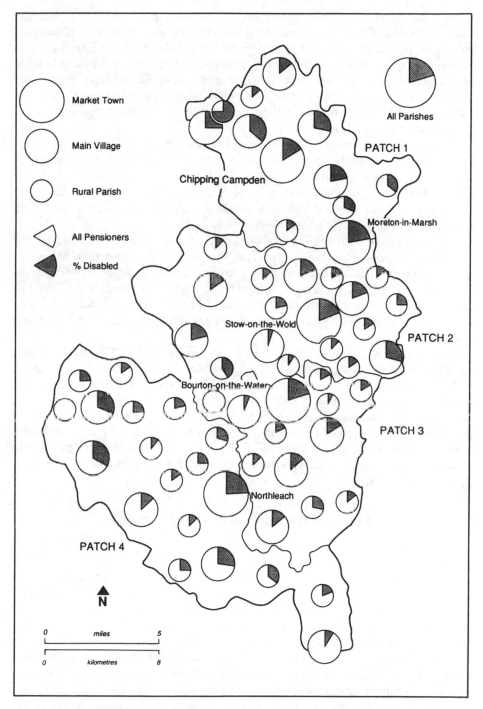

Figure 8.2   North Cotswolds: disabled pensioners as a proportion of all pensioners (Source: Gloucestershire County Council, 1986b, Parish Profiles, Table 9)

This procedure yielded information on 297 (73 per cent) of the target households with 498 elderly and/or disabled adults. These individuals comprised three groups: retired-elderly (76 per cent), retired-disabled (20 per cent) and younger disabled persons (4 per cent). Eighty-five per cent of the sample were aged over 60 and 31 per cent over 75. Included among the retired were a small minority of the early-retired who, for health and personal reasons, had left full-time employment a few years before the statutory retirement age of 65 years for men and 60 years for women.

Most of those enumerated in the sample used simple health aids, including spectacles (68 per cent), walking frames and sticks (11 per cent), hearing aids (6 per cent) or special appliances of various kinds (7 per cent). Almost 25 per cent were afflicted with multiple and more serious disabilities. Arthritis, the most common disability, affected one third, and often aggravated situations of failing eye-sight, deafness, angina and the immobility of limbs.

For the analysis of journey patterns, Table 8.1 groups these individuals into two kinds of household with respect to the nature and degree of infirmity. *Independent* households (72 per cent) contain pensioner members who are in reasonable health, including the lone-retired. *Dependent* households (28 per cent) include at least one member afflicted by some serious disability and who needs the regular help and support of others to cope with many activities essential to daily life.

This division was justified when individuals were scored on their performance against a standard and graded set of personal and domestic tasks (Bebbington, 1979). These involved the ability to use public transport and shops, movement inside and outside the home and the management of personal care and toilet arrangements. The elderly-independent had few problems in performing any of the specified tasks. Disabled persons, in contrast, found considerable difficulty with a number of these activities. Furthermore, disability restricts movement outside the home and, even with assistance, one third were unable to use a bus or visit the local shop.

Access to transport and travel patterns

Access to a household car and regular bus service are important determinants in the travel behaviour of those who can venture outside the home. These extramural movements are confined almost exclusively to shopping and personal service trips, visits to relations and friends and to places of worship and communal groups (Warnes, 1987). In the eight parishes 75 per cent of households own at least one car, but the proportions for independent (60 per cent) and dependent (56 per cent) households are appreciably lower. For many of these car driving is curtailed at night and restricted to familiar routes. More significantly, in relation to personal mobility, only 38 per cent of individuals hold a current driving licence and enjoy unrestricted access to a household car (Table 8.2). Consequently, the journeys made by large numbers of females, the retired-disabled and those aged over 75 are severely constrained (Hitchcock, 1980). These disadvantaged groups, moreover, are strongly represented in the

Table 8.1 Household type

Household type	Independent	Dependent	Total
Lone pensioner:			
Male	23 } 92	4 } 26	27 } 118
Female	69	22	91
Other pensioner:			
Households	122	44	166
Non-pensioner:			
Disabled	0	13	13
Total	214	83	297

(*Source*: Field Survey, 1982.)

12 per cent of households that, for reasons of health and finance, had relinquished the sole household car in the five years preceding the survey.

Throughout the region, therefore, many of the active elderly and disabled relied on limited public transport for their basic travel needs (Gant and Smith, 1985b). In the four weeks preceding the survey, the proportion of individuals using local bus services ranged from 44 per cent at Chedworth to 15 per cent in Southrop (Table 8.3). These differences and the varying proportions of the *independent* and *dependent* using public transport can be explained by personal circumstances, the frequency of bus services and the compensating quality of mobile services in the smaller villages. Not unexpectedly, a greater overall proportion of the independent (41 per cent) than the dependent elderly (25 per cent) had recently made bus journeys, the majority in each group taking advantage of bus tokens issued by the county council.

These disparities in personal access to transport were reflected in the trip patterns to four representative services: a grocer, post office, dispensing chemist and hospital (Figures 8.3–8.6). For this exercise based on personal travel behaviour in the previous month, individuals were divided into three categories: the 'housebound', including the severely handicapped, who remain indoors and depend on other persons; those satisfying their needs in the home parish; and the remainder who travelled farther afield.

For *grocery purchases* the substantial flows to Cheltenham, Charlton Kings and Cirencester were balanced in volume by localized movements to minor centres, such as Andoversford, Northleach, Burford, Witney, Carterton, Lechlade and Fairford (Figure 8.3). In all parishes significant proportions of both the *independent* and *dependent* were housebound and were forced to rely on friends, neighbours and relatives for shopping, while many others were restricted to using local (mobile) services. On balance, however, a greater proportion of the *independent* made trips outside the parish and this movement was observed even in those parishes such as Bibury and Coln St Aldwyns, which supported a village shop. The most striking local contrasts in movement

Table 8.2 Individual's access to private transport in relation to personal characteristics (%)

Availability of private transport	Retired, living alone	Retired, living with others	Retired and disabled	Under retirement age and disabled	Male	Female	Aged over 75 yrs	Total
Available: Car and licence	36	45	28	23	56	26	25	38
Partially available: Car/no licence	—	24	14	54	16	31	15	18
Not available: No household car	64	31	58	23	28	43	60	44
Total	100	100	100	100	100	100	100	100
N =	115	261	100	22	208	290	155	498

(*Source:* Field Survey, 1982.)

Table 8.3   Use of bus services in month preceding the survey

Parish	Independent		Dependent		Total samples	
	Persons	% using bus	Persons	% using bus	Persons	% using bus
Bibury	60	47	35	23	95	38
Chedworth	91	45	25	40	116	44
Coln St Aldwyns	16	6	16	38	32	22
Eastleach	30	47	11	18	41	39
Sevenhampton	49	49	15	13	64	41
Southrop	30	20	10	—	40	15
Whittington	18	27	9	33	27	30
Yanworth	8	50	4	—	12	33
Total	302	41	125	25	427	36

(*Source*: Field Survey, 1982.)

behaviour between the *independent* and *dependent* were found in Eastleach, Sevenhampton and Southrop.

Access to a *post office* is especially important to an old-age pensioner. Figure 8.4 shows that the majority of those interviewed used a local post office or one in a neighbouring village, although many of the *independent*, with private transport, combined their visit to a post office with shopping or social activities in a large centre. The *dependent*, in contrast, are characteristically more restricted in their movements and a far greater proportion needed help to collect a pension or to use ancillary services available at their local post office.

In most respects the patterns of journeys to a *dispensing chemist* shown in Figure 8.5 resemble those to a doctor's surgery. Not one of the eight parishes had a dispensing chemist; however, the doctors holding part-time surgeries in Chedworth and Bibury dispensed prescriptions. Moreover, the housebound and others were often fortunate in having car-owning members of the community to ferry them to doctors' surgeries and to collect prescriptions from chemists located in the minor centres.

Relatively few of either the *independent* or *dependent* were required to attend *hospital* in the month preceding the survey. Figure 8.6 shows, however, that visits to hospitals in the major centres like Cheltenham, Cirencester, Oxford and even London were complemented by more localized journeys to the cottage hospitals in Bourton-on-the-Water, Fairford and Northleach.

There are naturally differences in the levels of demand for these basic services. Table 8.4 indicates that, in the previous month, 63 per cent of the 498 people in the sample had visited a grocer, 59 per cent a post office, 42 per cent a dispensing chemist, 24 per cent a doctor's surgery, while only 6 per cent had been required to attend hospital. Travel frequencies varied according to the circumstances of these individuals: in the month preceding the survey, and on at least four occasions, 85 per cent had visited a grocer, 74 per cent a post office, 39

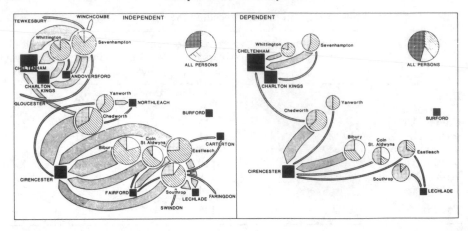

Figure 8.3   Trip patterns: journey to grocer (Source: Field Survey, 1982)

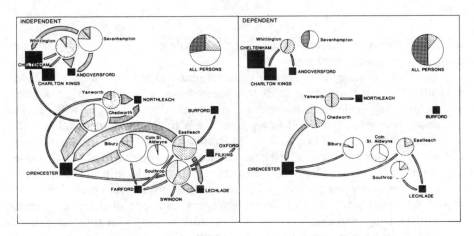

Figure 8.4   Trip patterns: journey to post office (Source: Field Survey, 1982)

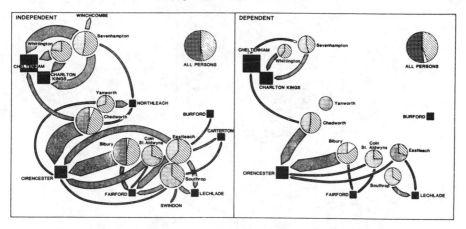

Figure 8.5   Trip patterns: journey to chemist (Source: Field Survey, 1982)

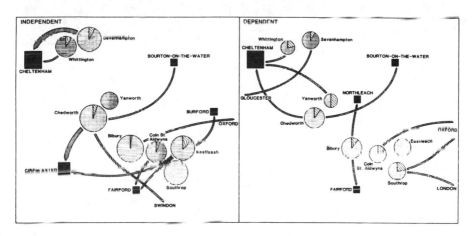

Figure 8.6   Trip patterns: journey to hospital (Source: Field Survey, 1982)

Table 8.4 Frequency of journey to selected services (based on the percentage of individuals making specified journeys)

No. of visits in previous month	Grocer			Chemist			Post office			Hospital			Doctor		
	Ind.	Dep.	Tot.	Ind.	Dep.	Tot.	Ind.	Dep.	Tot.	Ind.	Dep.	Tot.	Ind.	Dep.	Tot.
1	1	3	2	34	51	38	11	16	12	100	91	97	63	100	69
2	9	15	10	13	25	16	13	9	12	—	—	—	18	—	14
3	2	6	3	8	2	7	1	4	2	—	—	—	8	—	7
4	70	59	67	41	22	36	69	61	67	—	—	—	9	—	8
5+	18	17	18	4	—	3	6	10	7	—	9	3	2	—	2
Total persons (= 100%)	252	63	315	167	44	211	239	59	298	15	15	30	98	20	118

Notes Ind. Independent; Dep. Dependent; Tot. Total journeys.

(*Source:* Field Survey, 1982.)

per cent a chemist, 10 per cent a GP's surgery and 3 per cent a hospital for treatment.

In general these trips were destined for the nearest available facility, and walking is clearly an important mode of transport for the elderly and less severely disabled (Hillman and Whalley, 1979). For the car-less, bus journeys were more common for grocery shopping and trips to a chemist while a significant minority depended on friends, neighbours and relatives for lifts to a doctor's surgery and hospital (Table 8.5).

Welfare implications

Most of the elderly and disabled in the eight survey parishes led independent lives. For others, however, fading health, limited financial resources and the absence of a household car have restricted personal mobility. Many could not have remained in their own homes without the committed support of their family, community and local voluntary organizations, especially in winter, and only those with special problems had been directed to the social services for regular supervision (Shaw and Stockford, 1979). Under the provisions of the Chronically Sick and Disabled Persons Act 1970 the statutory social services have a crucial role to play in community care. There is, however, widespread under-reporting of disability in the eight survey parishes and in the four weeks preceding the survey only 19 households had benefited from the services of a home-help, 5 received meals-on-wheels and 9 had been visited by a social worker or occupational therapist.

Fortunately, family-support systems are well developed in the study parishes. For those without immediate kin, there are effective and compensatory systems of neighbour and community support (Harper, 1987c; Wegner, 1982, 1988). Frequent and caring contact has an important psychological benefit in reducing isolation; there are also more tangible benefits, such as help in times of stress and illness, a reassuring presence and the provision of a car for various shopping journeys and hospital visits, and direct assistance with heavy and demanding household tasks. Seventy-two per cent of households had close relatives living within 40 km of their homes; half had children and/or grandchildren living independently in the same or a neighbouring village. Frequent contact was the norm, usually by reciprocal visits, and only 10 per cent of the respondents had not seen at least one relative in the month preceding the survey. These systems of domiciliary support, however, are under serious threat. The total number of the elderly and infirm continues to rise; so, too, does the population without caring relatives living nearby. The predicted increases in the local cohorts aged over 75 who will suffer progressively greater limitations on personal health and restrictions on mobility will aggravate this situation (Alderson, 1988).

Reductions in local services, including public transport, have created serious problems for the elderly. As Figures 8.3–8.6 demonstrate, significant proportions of both the independent and dependent are unable to travel too far from the home parish; indeed, many are truly housebound, and some bedridden. As in many other rural areas, the Transport Act 1985 has not stemmed the decline

Table 8.5 Principal mode of travel and journey purpose (based on the principal mode of transport used for specified journeys) (%)

Mode	Grocer			Post office			Chemist			Hospital			Doctor		
	Ind.	Dep.	Tot.	Ind.	Dep.	Tot.	Ind.	Dep.	Tot.	Ind.	Dep.	Tot.	Ind.	Dep.	Tot.
Walk	9	19	12	52	56	53	2	—	2	—	—	—	17	5	15
Bus	28	19	24	11	2	9	23	34	27	22	—	13	15	5	14
Household car	54	56	55	31	32	31	63	57	61	33	50	40	55	60	56
Other car	9	6	7	5	10	6	12	9	10	39	42	40	13	30	15
Other, e.g. cycle	—	—	2	1	—	1	—	—	—	6	8	7	—	—	—
Total (100%)	252	63	315	239	59	298	167	44	211	18	12	30	98	20	118

Notes Ind. Independent; Dep. Dependent; Tot. Total journeys.

(*Source:* Field survey, 1982.)

of public transport but has heralded a further layer of policy-induced deprivation (Farrington, 1986). The deregulation of the bus services to stimulate competition, improve efficiency and reduce public subsidy has had serious consequences for the route network and frequency of services in the North Cotswolds. For decades local independent bus operators have provided life-line services to the mobile who lack private transport. In the past these small companies had knowingly cross-subsidized some operations to maintain uneconomic routes and Sunday services (Gant and Smith, 1985a). That situation was inevitably changed by the Transport Act 1985. Despite county council intervention to subsidize selected routes that were not maintained by market competition, the pre-1985 network contracted even further (Gant and Smith, 1985c). Notwithstanding the stability of the bus services since deregulation, the legislation has created serious consequences for the car-less elderly in the smaller and less accessible settlements that have already lost most of their basic services.

Community transport initiatives

Within the changing framework of government policy (discussed in Chapter 9), increasing attention has been focused on the voluntary sector for the provision of rural transport (Banister and Norton, 1988; Nutley, 1988, 1989). In rural Gloucestershire, however, community transport initiatives are piecemeal and offer no real alternative to regular, frequent, and wide-ranging public bus services responsive to the local needs of disadvantaged groups, particularly the elderly and disabled (Mayo, 1983; Cloke and Little, 1987). There are, however, two effective examples of community transport schemes operating in the Cotswold District, and these represent models for evaluation and possible replication: the Villager Community Minibus, based at Oddington near Stow-on-the-Wold, and the Voluntary Help Centre car-sharing scheme administered from Moreton-in-Marsh.

The Villager Community Bus

From a modest beginning in 1981 when the Villager Community Bus Service was launched to serve seven villages in the vicinity of Stow-on-the-Wold, it has expanded to a network of routes linking 60 communities with a number of market towns including Burford, Chipping Norton, Moreton-in-Marsh, Winchcombe and Witney. Passenger numbers have risen steadily from 1,452 in 1982 to 17,190 in 1989, making the Villager one of the most extensive community bus services in the country. The success of the operation may be attributed to a number of factors including a force of 15 volunteer drivers; support at parish, district and county level; and (not least) the chairman of the (voluntary) management committee who has been responsible for the day-to-day operation of the service since its inception in 1981. Financial independence has accompanied the expansion of the network. The original second-hand vehicle and its 1983 replacement were acquired with grants from Gloucestershire County Council. By 1988, however, profits from revenues allowed the purchase of a new 16-seat Ford

Transit, adapted with a wheelchair tail-lift and financed by the Rural Transport Development Fund. When not required for public transport duties both vehicles are available for hire by schools and other voluntary organizations.

Expansion of the route network subsequent to the Transport Act 1985 increased the number of villages served from 28 in 1985, to 57 in 1987 when the route mileage totalled 33,472 miles. With the exception of a fortnightly service to Winchcombe through five villages, however, new routes have centred on Witney and serve villages outside the North Cotswolds. In general, therefore, the services lost through deregulation in the south of the study area have not been compensated by a growth in community buses, thus placing additional burdens on other forms of community transport.

Voluntary Help Centre car-sharing scheme

As a consequence of the progressive reduction in the county council Rate Support Grant since 1984, the allocation to Gloucestershire ambulance and ambulance-car services has been cut. This has resulted in a sharp increase in the demand for community transport in the north Cotswolds. The Voluntary Help Centre at Moreton-in-Marsh has responded to this situation. Set up by local initiative in 1978 and staffed by volunteers, it covers the 63 parishes of the region. Its basic objective is to co-operate with the statutory and voluntary organizations in finding volunteers for community work, and to offer advice and services to those in need. Since its inception it has co-ordinated a number of schemes, which have benefited the elderly and disabled. These include the establishment of a network of village contacts, mobile information services, a gardening and home decorating service and a Deaf and Hard of Hearing Club.

Most important, however, the centre has acted as a catalyst for the voluntary car-sharing scheme. This was inaugurated in 1984, and depended on retired car-owners for its success. In 1988 there were 98 volunteer drivers on the register. These volunteers used their own cars to drive those in need to authorized destinations. Their passengers are of two kinds: category A – those people without their own transport and with mobility difficulties by virtue of being elderly and frail, handicapped or wheelchair bound, which makes travel by conventional public transport difficult, unsafe or impossible; and Category B – able-bodied people without cars who, by virtue of the lack of adequate public transport services, are unable to make essential journeys, for example, to job interviews (Gloucestershire County Council, 1985b). Gloucestershire County Council supports this initiative with legal advice, publicity and an operating grant in 1989 of 14 pence per mile. This sum, together with a near-equivalent contribution from the passenger, is used to reimburse the driver.

This community transport initiative has been an outstanding success. In the period 1984–5 to 1987–8 the number of journeys facilitated by the Voluntary Help Centre has increased from 510 to 1,508, passengers carried from 1,800 to 3,477 and total mileage from 19,200 to 46,000 (North Cotswolds Voluntary Help Centre, 1984–8). Figure 8.7 shows that in each month in 1987–8 between 100 and 160 return journeys were organized and executed, the majority for the frail

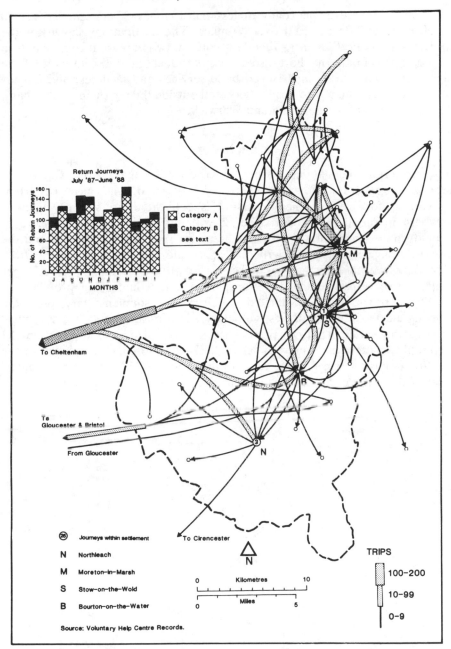

Figure 8.7   Car journeys arranged by the Moreton-in-Marsh Voluntary Help Centre, January to June 1988 (Source: Voluntary Help Centre records)

and elderly. Of these, 37 per cent were less than 10 miles, 30 per cent between 10 and 30 miles and 33 per cent over 30 miles. The destinations shown for the representative period January–June 1988 confirm two important components in the geographical pattern: short-distance visits to local market centres from the neighbouring parishes, which lacked basic services and facilities; and longer-distance journeys, mainly to hospitals located outside the region in Cheltenham, Gloucester, Bristol, Leamington and Warwick.

Conclusion

These are serious inequalities in access to basic services in the North Cotswolds; these are exaggerated for the elderly and disabled, and have important implications for welfare. This situation is replicated, to varying degrees, elsewhere in the British countryside. It will not diminish. The pressure of demand on environmentally attractive areas for retirement migration and the increased longevity of the elderly will aggravate the problems arising from inadequate levels of basic services and statutory provision. Imaginative schemes of alternative transport are, therefore, needed to counter the problems of mobility deprivation. Co-ordinated initiatives in supplementary provision involving elected representatives, voluntary agencies and public services must relate to the varied needs of the elderly and disabled, and respond to the geographical pattern and timing of demand. The development of policies for the retention of basic services within reach of the mobile elderly should be a central thrust of this strategy and a defined objective of the planning process.

9

PUBLIC TRANSPORT IN THE COUNTRYSIDE: THE EFFECTS OF BUS DEREGULATION IN RURAL WALES

Philip Bell and Paul Cloke

The advent of the Thatcher government in 1979 ushered in a great many changes to the economy and society of Britain. These changes have been of such impact and magnitude that 'Thatcherism' has been identified as a distinct movement, and its acolytes are especially keen to claim not only that the rest of the world has come to follow the UK's lead but also that policy developments under Thatcher have been entirely beneficial. More detached opinion has varied, however, with some identifying a carefully planned programme of policies while others characterize the whole process as operating largely by chance and opportunism. Pure chance it cannot be, however, as similar trends have also been occurring in other nations, not just Western capitalist ones but also percolating into the former command economies of Eastern Europe and various Third World nations.

This chapter sketches the growth of privatization, perhaps the main identifiable element of the New Right thinking in practice, and disentangles some of the initiatives that combine under this heading. It then turns to consider the effects of privatization on one particular sector, that of public transport, which is illustrated in detail by a case study of bus deregulation in rural Wales, and concludes by drawing out some of the main themes in order to assess the likely impact of privatization upon rural Britain.

Privatization and the New Right

The New Right movement may have achieved recent prominence, but many of the core ideas it espouses have existed on the fringes of political activity for some time, most notably in the free-market ideals espoused by Hayek and Friedmann. Proponents of these ideas have captured the policy process as a reaction to the perceived failure of previous attempts to control the economy, as represented by the weaknesses in the Morrisonian model of nationalization and the growing crisis in funding the Welfare State, and have been supported by powerful economic interests. As they translate into reality, however, they move from being an apparently flawless blueprint able to sweep aside the practices they aim to replace, to becoming a complex process whose various negative effects have become increasingly obvious and have generated opposition from many, if relatively powerless, quarters. Political reaction has grown recently, and it is quite possible that many of these initiatives will be reversed in future. However, many commentators who are basically opposed to the process have been prepared to concede that previous practice was far from perfect, and that new approaches have to be developed (Le Grand, 1983; Davison, 1985). If privatization is the wrong answer, at least it may stimulate more productive thinking in response, and allow progressive forces to recapture the practical and theoretical high ground.

In the space available in this chapter it is impossible to do full justice to the range of New Right ideas and resulting critiques; a supportive review is provided by Green (1987) and a more critical analysis by King (1987). It should be noted here, however, that the body of New Right thought is riven by a basic tension, which has not yet proved terminal but has given rise to alternative views on privatization and its effects. New Right *liberals* have placed particular emphasis on total freedom in the economic sphere, with the role of the State minimized, if not removed altogether, and with most regulation being self-imposed. More *conservative* elements might concur with the objective of reducing the influence of the Welfare State, which is seen as detrimental to the spirits of self-help and enterprise, but are worried about the possible moral effects of a free market, as witnessed for instance in the debates over the degree of regulation that is desirable in broadcasting. Harmonizing these two wings to some degree therefore requires the development of a formula such as a free economy within a strong state (Gamble, 1988), and can never be entirely satisfactory. In addition, whilst fragmented individual, civil and political rights have been elevated, social, collective and political rights have been attacked (Dunleavy and O'Leary, 1987). New Right ideas thus contain a number of potential fault lines, which could form lines of attack as the general approach begins to fall from favour.

In practical terms, privatization has been carried forward in a number of different arenas, and several classifications have been devised to cope with this complexity (Lundqvist, 1988). Heald's (1984) fourfold division gives some order to the policy field, although it is somewhat coarser than other schemes. He distinguishes the following:

1. *Charging* The increasing tendency to charge for goods at market prices, and the removal of subsidies from central or local government. This often accentuates previous practice, rather than instigating a new approach.
2. *Contracting out* The compulsion to put a variety of (mostly local government and health) services (e.g. street cleaning and rubbish collection, school-meal provision) out to competitive tender. This ideally brings in private firms to operate at lower cost, but can also force in-house firms to make wage or job cuts to retain work (Ascher, 1987).

3. *Denationalization* The transfer or return to the private sector of many nationalized industries, often in part at first, but now with most largely divested of State control. This is not only seen as making the industries more efficient but also as being electorally popular, as shares are made available at heavily discounted prices, allowing rapid windfall gains. Questions are raised, however, over such issues as the loss of control over important sectors of the national economy, asset stripping and resultant redundancies.
4. *Deregulation* Many markets have been opened up with previous rules restricting free competition being abolished or relaxed. However, in several cases denationalization has been accompanied by some increased regulation to control the deleterious effects of a free-market operation, as when new regulatory bodies have been installed, such as the National Rivers Authority for the water industry or OFTEL for British Telecom (Veljanovski, 1987).

These four strands can complement each other, as when raising prices or removing barriers to entry precedes denationalization, but conflicts can also be apparent. Early sales of public industries largely took the form of privatizing monopolies intact (e.g. British Gas), so little effective competition resulted. Political embarrassment has forced a less monolithic approach subsequently, as in the electricity and water industries. However, the above classification does provide a useful framework for an examination of changing policies in the transport sector and it is to this we now turn.

Control of public transport

In the transport sector itself, various regulatory devices can be traced back well into history. The first widespread and real example is perhaps provided by the railway network. In its development in the last century, competition was the watchword, with Parliament very concerned to prevent collusion and monopoly as many minor railways were taken over and significant groupings formed. Several proposals for amalgamation were rejected, not just where area monopolies would have developed but also in end-on schemes where it was hard to see that competition would be affected; challengers might well have been made more efficient by such schemes. It may be doubted whether such provision was wasteful in social terms, but certainly many later developments were at best marginal economically, and the network at its height greatly exceeded what would have been provided under more regulated conditions. Eventually, under pressure from new road-based competitors, and in some cases declining markets, the railways were gathered into four major groupings in 1923, and these in turn became the single British Railways in 1948. Economies of scale and integrated planning formed a major part of the justification, coupled with the parlous state of the railways after operating under the strain of the war years. Before passing on, however, it should be noted that competition was still insufficient to provide services to many remoter areas. Towards the end of the century, the government relaxed many provisions on quality in encouraging light railways, and when even this proved insufficient in areas like the west of Ireland, made quite specific contributions, along with local authorities, to assist in their construction (Salveson, 1989).

In the bus industry, rapid expansion took place in the years after the First World War within a relatively unconstrained framework. This growth occurred in parallel with increases in private road-freight haulage, often based on the use of ex-military vehicles. At the time, the most significant division was between road and rail, but over the years this has shifted to a public–private division as the bus interest has come to parallel that of the railways more closely. Very fierce competition occurred on the roads in the early stages, often with worrying consequences for safety (Glaister and Mulley, 1983). Unease over this aspect in particular led to the Road Traffic Act 1930, which was a major landmark in the regulation of the bus industry. Despite the traditional wisdom that the main instigators for this legislation were the railway and tramway interests, which were still powerful enough to place some shackles on a menacing competitor, Mulley's (1983) investigation of the written sources dating from that time gives a different impression. In this version of events, safety questions were the dominant consideration, but in recognition of the financial burdens likely to be placed upon the operators in order to raise standards in this respect, regulation was introduced to protect existing firms from competition and resultant further loss of revenue.

The Traffic Commissioners were set up to oversee the industry, with jurisdiction over both quality aspects (each vehicle having to carry a certificate of fitness) and quantity (each route needing to be approved by the commissioners, who would take into account road capacity, existing provision and objections from the railways and other bus operators). Once this legislation was in place, concentration took place in the industry, with amalgamations and takeovers producing fewer but larger firms, and the network extending into more remote areas and threatening the rail companies.

For a short period after 1945, all transport sectors were brought together under the British Transport Commission, and some attempts at co-ordination were made. Denationalization of road haulage under the Conservatives after 1951 instigated the fragmentation of transport planning, and after the easing of post-war travel restrictions, both rail and bus services came under increasing pressure, especially in rural areas. The rural rail network was savagely pruned, particularly as a result of the 1963 Beeching Report, whilst at the same time the Jack and Aaron reports investigated the bus network. Service withdrawals and cutbacks were the end-result of a process whereby many routes moved into deficit, and the burden of cross-subsidy fell onto fewer profitable services – a problem increasingly apparent in urban as well as rural areas. As in the railways twenty years previously, nationalization was thought to be an appropriate response, and under the Transport Act 1968 the largest surviving operators were formed into the National Bus Company (or the Scottish Bus Company north of the border). Public transport for the conurbations came under the control of their passenger transport authorities and their implementing bodies, the executives. Plan-making and financial provision for public transport was addressed more specifically, with the Transport Policy and Programme (TPP) system followed by Public Transport Plans under the Transport Act 1978. The hallmark of the approach to transport up to 1980 was therefore similar to that in many other sectors: comprehensive planning, restriction of competition and a search for economies of scale.

This broad regulatory approach was reversed dramatically after the Conservatives returned to office in 1979. Transport was one of the few areas for which privatization proposals had been prepared at that time, and the 1980 Act contained proposals to deregulate the express-coach sector almost entirely with only some safety measures remaining (Kilvington and Cross, 1986). Initially, this gave rise to an explosion of fierce competition but this was not maintained and it seldom happened on more rural routes. Indeed, many settlements lost their services as schedules were accelerated and bypasses were used.

Regulations were relaxed at the same time in the stage-carriage market, and some early cases of competition have been documented (Savage, 1985). More significantly, county councils were invited to nominate trial areas where experiments could be conducted into the effects of a more complete rolling back of existing rules. Despite such blandishments, only Norfolk, Devon and Hereford & Worcester nominated parts of their area for such treatment, and in the first two very few significant developments resulted. Proponents therefore argued that existing regulations could be removed without serious adverse effect, while critics drew the lesson that the existing legislation was not a particularly significant factor preventing a revival of the market.

The results in Hereford & Worcester were more notable, and several commentators have documented the history of this initiative (Dunbar, 1984; Jones, 1986; Evans, 1988). Most changes took place in the city of Hereford itself, where competition was initially fierce, and many routes saw reduced fares, but there were also severe problems over vehicle safety and passenger uncertainty, as major changes in services continued for far longer than was expected. The incumbent National Bus Company, Midland Red, after losing out initially, responded to competition and regained much of its market. Changes in the rural areas around Hereford were far less drastic, with little competition but few actual route losses (although some occurred, linked to competition in the city). The county made savings on its subsidy bill, but seemed to accept that Hereford was in many ways atypical, having a high number of private competitors waiting in the wings. Hereford then provided few conclusive answers, but was used to justify subsequent legislation to extend this approach to most of the country (although not quite in the form propounded by Hereford & Worcester). This provides an illustration of the 'privatization as theology' approach commented on by Heald and Thomas (1986).

A white paper in 1984 was followed by a further Transport Act in 1985, under which privatization was taken much further. While *quality* licensing was maintained, *quantity* controls were removed, so that operators were largely free to run any route they wished, and could no longer be blocked by objections by other operators. Routes or services not operated commercially could be subsidized as 'socially necessary' by local authorities, and were to be subject to competitive tendering, and the National Bus Company (NBC) was to be broken up and sold off; only London was to be exempt at first (Higginson, 1984).

The overall approach combined all of Heald's four main elements of privatization, as not only was the market to be deregulated significantly but the NBC itself was also to be fragmented and privatized, tendering was to replicate contracting out by introducing many small private operators and subsidies were to be reduced and many fares were to rise, especially in the conurbations where authorities such as the Greater London and Sheffield City Councils had operated cheap-fares policies to encourage passenger travel. Indeed, it could be suggested that the desire to control such authorities was a prime motivation for the overall bill, and its implications for rural areas were barely considered as it was processed to become legislation.

Bus deregulation in rural Wales

Basic trends and patterns

Once the legislation had been implemented, the key issue became one of assessing the impact of privatization in practice. The present authors have already argued for empirical studies to cast light on the actual implications of various privatization measures for rural residents (Bell and Cloke, 1989), and work on a research project on bus deregulation in mid and north Wales for the Department of Transport and the Welsh Office (Bell and Cloke, 1988) enables some conclusions to be reached in this respect. The study area is a particularly remote one in English and Welsh terms at least, and so cannot be taken uncritically as representative of other rural areas, but the basic trends seem to have been replicated elsewhere in a recognizable form.

The legislation required all operators wishing to run commercial services to register these with the Traffic Commissioners by March 1986, so that the county councils had sufficient time before the start of deregulation to decide which other services they felt to be necessary, operate a tendering process and award contracts. The new system began on 26 October 1986, and all operators had to run registered services for three months; after this, they could vary, add or cancel services at 42 days' notice, thus making the full deregulated framework very flexible. Counties had to react very quickly to any network changes, although operators still complained that they could be tied to an unprofitable situation for too long! The situation in the study areas of West Clwyd (Rhuddlan, Colwyn and Glyndwr districts) and North Powys (Montgomery district) can only be summarized in this context; fuller details are provided elsewhere (Bell and Cloke, 1988).

The base network (that existing in November 1985 at the commencement of the study period) showed North Powys to be dominated largely by the local NBC company, Crosville (Figure 9.1). They ran the main, regular routes (notably Shrewsbury–Welshpool–Newtown–Llanidloes, Welshpool–Oswestry and Machynlleth–Newtown) and several feeders. Several, smaller companies

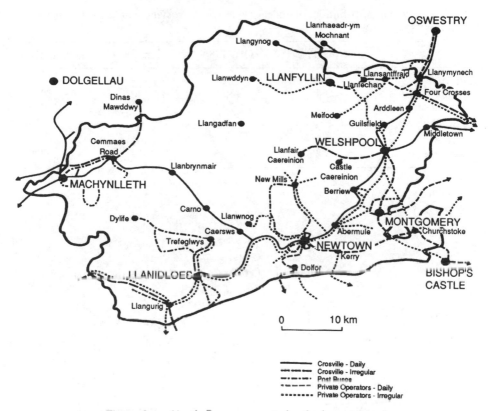

Figure 9.1   North Powys case study: the base network

did operate in the district, however, especially in the east, providing a reasonable pool from which competitors might be drawn. Many were involved in schools' work, though only a few ran regular, daily stage services; the Oswestry–Llanfyllin route was a rare example. Four post buses also operated in the area, two from Llanidloes and the others from Machynlleth and Newtown. Overall, Crosville began with 34 per cent of route mileage, but 62 per cent of vehicle miles per week, confirming that they ran the more intensive routes on the main corridors (Table 9.1). Even so, some restructuring had already taken place, and more took place in the run-up to deregulation, involving service reductions and (in Newtown) withdrawals.

The registered commercial network proved to be very sparse (Figure 9.2). Although half the route mileage was registered, this was perhaps inflated by some long-distance, limited-stop services; and only about a quarter of vehicle mileage was judged to be commercial. This gives a broad picture of the network potentially at risk from any reductions in subsidy. Crosville registered only half of their journeys on the Shrewsbury–Newtown route, and nothing on their other major services. This left the county with a major problem in maintaining a

Table 9.1 Overall distribution of mileage between Crosville and private operators (north Powys) (%)

	Nov. 1985	Oct. 1986	Jan. 1987	Jan. 1988
Route mileage				
Crosville	34	30	30	30
Private operators	66	70	70	70
Bus miles/week				
Crosville	62	61	61	63
Private operators	38	39	39	37

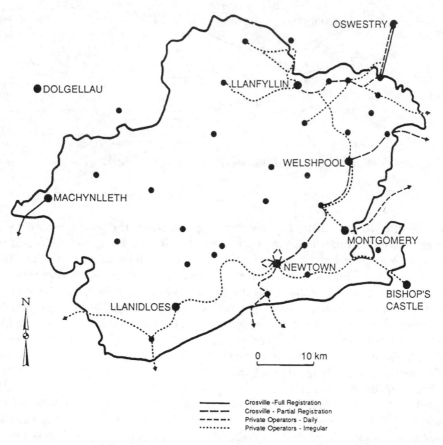

Figure 9.2   North Powys case study: the commercial network

network after deregulation, but ultimately they were able to replace services at approximately the same cost. In practice, Crosville proved very successful at retaining services under contract and, although most tenders attracted at least one competitor, very few changed hands, even between some of the private operators. The most notable Crosville losses were that of the Tanat Valley service from Llangynog to Oswestry and the two-day-a-week journey between Welshpool and Llanfair Caereinion. The results of this feverish activity were that on commencement of the deregulated system in October, route mileage had fallen overall by less than 3 per cent, although vehicle mile losses exceeded 11 per cent.

Subsequent changes proved to be minimal over the period of study. A small number of extra services appeared, some of them being county council sponsored, so by the beginning of 1988, losses in mileage relative to the base level had been reduced from 2.6 to 1.6 per cent (route miles) and 11.3 to 7 per cent (vehicle miles). The major events in Powys were one private company requesting and receiving subsidy for routes initially registered as commercial, and a small instance of competition in Newtown itself. The first of these did not involve substantial sums of money, and was agreed without going through a process of retendering, but would obviously be a worry to Powys if repeated on a wider scale. Overall, then, Powys' network had changed little when the dust of deregulation had settled. Crosville were still dominant, losing some marginal services, but with their share of vehicle mileage little altered.

The situation in west Clwyd was rather more complex, although the basic conclusions are broadly similar. Here, a fundamental division exists between the populous north coast (from Colwyn Bay to Rhyl and Prestatyn), and the rural hinterland. Again, Crosville dominated the original pattern, particularly the north, with private services being most numerous around Wrexham (Figure 9.3). They ran two thirds of route mileage, and 90 per cent of vehicle miles (Table 9.2). No post buses ran, but a community bus operated out of the Uwchaled area around Cerrig-y-Drudion. The commercial network proved to be very concentrated, confined to the north coast (plus the main inland Rhyl–Denbigh route) and the south east extending to Llangollen and Oswestry (Figure 9.4). However, in Clwyd the commercial vehicle mileage was higher than route mileage (53 to 39 per cent initially), as most of the busy north-coast routes were registered, except at the service margins. (In Powys it was the sporadic market-day journeys that had predominated, often relatively high on mileage but low on frequency.) This pattern again illustrates the absolute reliance of most rural areas on public subsidy. Independents made rather more inroads into the market in Clwyd as a result of tendering, cutting Crosville's share of route mileage from 66 to 51 per cent, and vehicle mileage less dramatically from 89 to 79 per cent. The only commercial challenge initially came in Colwyn Bay, where an independent operator fulfilled a longstanding idea of competing with the NBC company, registering two competitive services. Route mileage fell by nearly 8 per cent on deregulation, but vehicle miles were largely unchanged; extra services instigated in January cut the route mileage loss to 5.6 per cent and added nearly 8 per cent to vehicle miles.

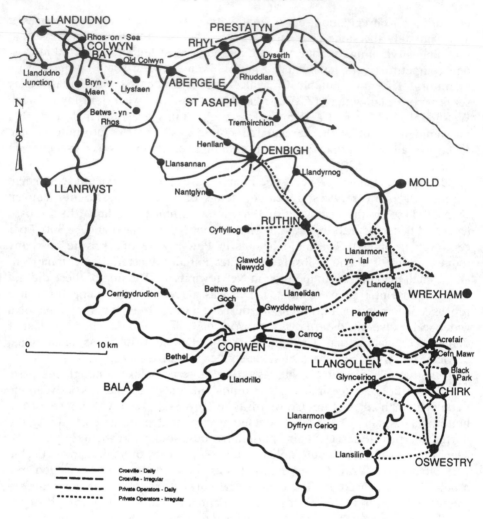

Figure 9.3   West Clwyd case study: the base network

Table 9.2   Overall distribution of mileage between Crosville and private operators (west Clwyd) (%)

	Nov. 1985	Oct. 1986	Jan. 1987	Jan. 1988
Route mileage				
Crosville	66	51	52	51
Private operators	34	49	48	49
Bus miles/week				
Crosville	89	79	82	82
Private operators	11	21	18	18

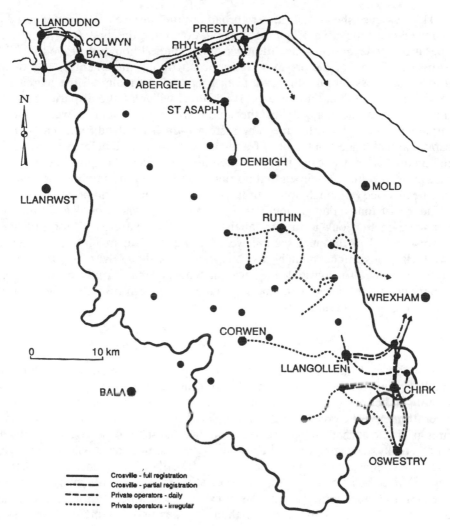

Figure 9.4 West Clwyd case study: the commercial network

Rather greater activity was visible once the full deregulated system was in place. By far the most significant factor proved to be the introduction of minibus operations by Crosville. They spread eastwards along the coast after an initial appearance in Conwy in late 1986. Many routes were restructured, and usually service frequencies were increased, this accounting almost entirely for the ultimate increase in vehicle mile figures by 30 per cent at the end of the study period. Minibuses percolated more slowly into rural areas, and here usually just replaced standard buses on the same schedules; travel opportunities therefore were not widened to the same extent.

This evidence shows that, in very broad outline, service patterns appeared to change very little as a result of deregulation over the first full year of its operation. Independent innovation was negligible, although it proved possible to maintain most of the pre-existing services without implementing any of the priority schemes for cuts that had been prepared. Virtually all the increase in service levels came in the north-coast corridor, widening the disparity between this zone and the majority of the study area. The counties were able to implement and administer the new system (which involved extra costs, both capital and revenue), but a lot of frantic activity took place below water to belie the picture of relative calm that prevailed on the surface. For proponents of the measures, it could be claimed that no routes had been lost, and in Clwyd money had been saved (although not in rural areas); opponents would point to the scant evidence of innovation or increase in opportunity, and would also question financing in the long term. An analysis of broad patterns is not sufficient, however, and in the next section we examine the two main local influences producing this effect, namely the bus operators themselves and the county councils. We also consider the bus passengers, those rural residents who experienced the effects of this new system, and in turn responded to them by increasing or decreasing usage.

The actors and agencies involved

The various privatization measures were supposed to enhance the ability of smaller firms to compete in the market, thereby galvanizing what were seen as remote and bureaucratic NBC companies into a more dynamic approach to their operations. As the preceding discussion has demonstrated, very few have been prepared to risk challenging Crosville, and most of the prospective new entrants who actually decided to enter the market have chosen to do so via the tendering route. Extreme caution has been the watchword, with defensive commercial registrations being made to preserve existing business, and many operators seeming to tender for each other's services more as a precaution to counteract any losses they might suffer rather than as a genuine statement of aggressive intent. Once these subsidized services had been gained, most were run to old schedules. One rare exception was Tanat Valley Motors, the new operators of the Oswestry–Llangynog route, who maintained that they were paying more attention to the actual needs of the valley communities and were reaping the benefits in terms of passenger numbers. Towards the end of the study period, extra journeys appeared in the schedule, giving later journeys back from Oswestry (which were subsidized by the county council). Other ideas were to strengthen the service up the more remote western end of the Tanat Valley, and to provide more unconventional services to link in with the main valley route. A negative factor in operation here, and perhaps elsewhere as well, was a fear of losing the route in a subsequent round of tendering if the service had been made to appear more attractive. This instance, however, was very much the exception proving the general rule. Most operators perceived rural Welsh territory to be infertile ground for ambitious experimentation, a factor that seemed to be repeated elsewhere in Wales.

Changes within Crosville proved to be far more significant. In the approach to deregulation, the company was split into English and Welsh sections. Initial commercial registrations had been made from the old company headquarters at Chester, but all subsequent operations had been managed by the new company Crosville Wales, based at Llandudno Junction. (All references to Crosville in the foregoing relate to the latter company.) As part of the selling off of NBC companies, Crosville Wales was purchased by management buy-out, although it has recently been sold again as part of the widespread concentration of capital in the industry. Certainly the new company appeared far more locally aware and innovative, and responded rapidly to initiatives from others. Savings were made through tighter schedules (which gave rise to some operational problems), redundancies amongst drivers and support staff, and lower wages (rates being disconnected from the former Crosville standard based on conditions in Merseyside). Owing to the increase in minibuses, employee numbers eventually recovered to and surpassed the previous level, but at lower rates of pay. Some asset realization has also taken place, out garages (such as Machynlleth) being sold, as was the bus station at Rhyl, the largest in the study area. Coupled with the increase in minibuses, including special seafront summer services, this threatened to worsen significantly congestion on the streets of Rhyl during the peak season, a matter of concern to Clwyd County Council.

Crosville also aggressively defended themselves against competition. The main challenge came around Wrexham, where fierce competition in the town ultimately spilled over into the study area around Llangollen. In this case, the main competitor was Wrights of Wrexham, a relatively large company with considerable experience of local passenger services. An intense, but brief, period of competition ensued, with rapidly changing routes and registrations, and causing great confusion amongst potential passengers. This ceased rather suddenly at the close of the study period, suggesting that a tacit agreement had been reached under which Wrights took over the main Wrexham–Llangollen service, but abandoned their attempts to displace Crosville from the rather more significant Wrexham–Chester route. Some inroad therefore was made, but even in the case of a strong competitor like Wrights, any damage to Crosville was limited. Overall, the latter company defended its main markets vigorously but was a little more relaxed about marginal rural business, being keen to retain uneconomical services but only if a reasonable return was available for doing so. Crosville Wales therefore retreated into its most profitable areas but yet retained many rural services, as the maps and figures demonstrate.

A heavy burden was placed on the two county councils involved in implementing the new system. Services have been more volatile than was generally predicted, requiring constant vigilance in filling gaps that might appear, or withdrawing contracts if a commercial service was instigated (as the legislation required). They have gradually gained more experience in operating the new system, and are able to take a rather more considered view of any situations now – a luxury not available to them in the mad scramble to reassemble a network for October 1986. Particularly in rural areas, contracts are usually stable once issued, running for periods of three years in most cases, and can often be extended.

Major initiatives are also underway in marketing and publicity, although some firms have remained aloof and preferred to issue their own (notably Crosville). All counties have attempted to issue co-ordinated timetables, although the general early rush, rapid service changes and the problematic commercial/subsidized split have all made this difficult to achieve. Confusion was widespread, particularly in the early days, and persisted long in certain places such as Llangollen, where patronage was clearly affected.

The final important consideration is that of the passengers themselves. In general services in many rural areas were little changed in terms of basic presence, cost and frequency, and therefore it is hardly surprising that passenger figures should appear little altered. Financial information provided to the county councils, and passenger surveys carried out by the authors in Rhyl, Denbigh, Llangollen, Welshpool and Newtown, confirmed that this was indeed the case. In Llangollen, however, as already mentioned, rapid service changes and incomplete information had led to a significant loss of confidence in the local service, a factor still reflected in patronage figures for the first year after deregulation. Rural bus usage has clearly not revived as a result of deregulation.

The north-coast area presents a slightly different pattern. Here, the introduction of minibuses has increased patronage, both generating more journeys from existing users and attracting more customers; the increased frequency is of significance here. Overcrowding and access can still be problems, however. Some initial fare reduction did take place in Colwyn, and later on in other places such as Rhyl, where competition was also possible. However, as in the rest of the country, fare reductions have been the exception not the rule. Crosville in particular have sought latterly to tie passengers to their own services by concentrating their discounts on weekly and monthly savers rather than individual journeys. What benefits there have been to passengers have therefore been concentrated almost exclusively away from the deeper rural areas.

The future prospects for rural Welsh public transport

The initial impacts of deregulation on the transport network in the study area were generally muted. As already stated, defenders of the measures could point out that, thanks particularly to the work of the county councils, deregulation had not proved to be the complete disaster predicted by some, though there were good political reasons for suggesting that it would not have been allowed to appear so immediately. Equally, however, any revival or reversal of the bus industry's fortunes was exceptionally well hidden. Many changes could be expected to take longer to work through, and could not be analysed fully within the remit of the research reported here. However, even if impacts take longer to emerge, they are no less a product of deregulation and should be considered as such. Two in particular are worth signposting at this point.

The first, crucial question concerns finance. The loss of cross-subsidy for rural operations was to be counteracted, the government claimed, by three factors. Competitive tendering would generate savings that could be used to bolster unremunerative services; a transitional grant for rural areas would be provided, tapering away over four years; and a specific grant would be made available for innovative services. In practice, Powys made no savings at all in the process, small gains in places being nullified by small losses elsewhere. Also, hardly any innovations were made, so the long-term prognosis when the transitional grant is phased out must be worrying. The county council suspects that larger operators have already allowed for this, but is less certain about smaller firms. If services are to be maintained at present levels, it seems likely that future tendering costs will rise. Clwyd County Council have made savings, though largely in the east of the county; very little of this can be attributed to the most rural areas. A contingency fund had been used, but not exhausted, in secondary tendering rounds, though a significant part of this saving appeared to be due to the county claiming the transitional grant for itself on subsidized services. Again, innovation was minimal, one scheme serving Bodelwyddan Hospital in particular failing to get off the ground. Familiarity with the system may enable extra savings to be made, but certainly the evidence from rural Wales suggests that the financial question is one that casts a long dark shadow over rural bus operations. Deregulation has not turned around the fortunes of rural transport and, unless the current level of commitment to provide facilities for the most disadvantaged rural groups is abandoned, subsidy costs must rise. Basic social inequalities are the underlying issue, not the legislative framework for transport. However, as the costs of support for individual routes are now clearer, county councillors may be less inhibited in suggesting cuts. There is also the risk (or hope for some) that the counties will be forced to take these unpopular decisions as budgetary constraints continue to tighten. Were this to happen the local authorities would be blamed accordingly, rather than attention being diverted towards central government, who are much more directly responsible for imposing these constraints.

Second, much was made of the role of privatization and deregulation in promoting competition. As has been shown, the scope for competition in the market has proved very limited in rural areas, although some has been evident over subsidized services, and may have acted as some restraint on tender prices. However, a most striking trend is the degree of concentration developing in the ownership of the privatized bus industry. Crosville Wales have recently been bought out by a freestanding, diversifying National Express group, and the emergence of super-companies such as 'Stagecoach' and 'Drawlane' is proceeding apace elsewhere.

The vision of many small competitors is thus being displaced rapidly by a return to a highly oligopolistic market structure, with possible tacit agreements against trespassing on each other's territory. This will also tend to reduce a county's options and cause subsidy bills to rise. There has already been growing evidence of careful manipulation of the timetables emerging in Wales as elsewhere, to gain subsidies for specific holiday periods.

In conclusion, therefore, a number of long-term concerns must be voiced over the effects of bus deregulation in rural Wales. Some improvements can be identified but overall the situation appears little changed, and once the carefully prepared transitional arrangements are peeled away, many of the longer-term trends are unfavourable. While proponents of the measures have some grounds for claiming that deregulation has not been instantly disastrous, less is heard of the claims for revitalization, which echoed through the original white paper and the debates around it. Certain positive features could be used in other frameworks, but overall it has done little for rural residents, and much wider measures would be required to increase opportunities here. These conclusions are echoed by the results of other studies (Bell and Cloke, 1990).

Privatization in rural areas

We now return to the question as to what extent these broad trends exhibited by Welsh rural public transport can be transferred to other areas and other sectors. Most attention will be focused on the latter, though it should be noted that rural areas elsewhere seem to have experienced few route losses, but equally little innovation and financial saving. The trade-off in urban areas between subsidy and fares is far clearer, the former's decrease leading to a rise in the latter, and although there have been more innovations (particularly involving minibuses), passenger numbers seem to have fallen (Tyson, 1988). Financial savings have been made in this sector but have caused additional costs elsewhere. Rail patronage revived due to the instability of the bus network, but the real concern is the transfer of people to cars, increasing congestion and leading to pressure for more road building. It would take a far more planned and co-ordinated system to change this particular emphasis, not to mention radical changes in governmental and civil-service attitudes.

Of greater relevance in this particular context is the effect of privatization initiatives on rural areas. It should be remembered, of course, that the impacts of privatization are very different for different groups, and therefore a simple urban–rural division is far too crude for the purposes of such an analysis. Privatization tends to benefit particular groups fairly consistently, and so the wealthier rural residents are likely to enjoy advantages denied to their poorer neighbours whose reliance on local services, which are under financial pressure, is far greater. On issues like share-ownership, despite the much vaunted claims about widening the interest for this, the benefits will flow to the wealthiest whose holdings are largest, while the loss of a potential long-term revenue source will hit those most dependent on government benefits. Similarly, contracting out of bus services might reduce the subsidy bill but will lead to redundancies or lower wages for bus workers, a group already far from wealthy, and thus depress their spending power on other services. Lower fares would benefit regular bus users, who are often less affluent rural residents, but this effect would be offset by a general reduction in network support, leading to service cuts and higher fares. The variety of initiatives under the privatization barrier will, overall, hit hardest those who are least able to withstand further financial pressures.

When considering rural areas in terms of their interests as locations of production, agriculture is the most significant and particular rural sector likely to be affected by privatization, principally in terms of subsidy reduction, but with some deregulation as well (perhaps indirectly through reduced planning controls over alternative enterprises). Many commentators have called for a less protected market that will reveal the virtues of competition by making farmers more self-reliant and thereby lead to a reduction in necessary government support (Howarth, 1985) – an approach already tried in New Zealand. Rural areas have also been affected by the privatization of other industries, although there is a far less pronounced rural dimension to these cases. More directly related to the public transport issue is the question of competition. As the transport case study demonstrated, very little effective competition, with the attendant benefits claimed in terms of efficiency, choice and price, arose in rural areas, and there are good reasons for expecting this to be repeated in other sectors. For services such as telecommunications, for example, it appears highly unlikely that rival networks will be provided to compete with the privatized giant of British Telecom, and so deregulation appears likely to provide few benefits for rural residents; the opportunity for profitable investment in such areas seems highly restricted. The loss of transitional funds may mean that cuts are equally severe, but just delayed to a period when it could be argued that it was not a direct result of privatization. In addition, unless some degree of social regulation remains, services may be withdrawn unless profitable, or not provided at all, particularly in the case of facilities with a high fixed component (for example, water and gas).

Even where services do remain, they will be vulnerable to the withdrawal of subsidy, and consequently higher prices under a user-pays policy. A commitment to uniform charging regardless of costs of provision for services such as the post and gas is highly unlikely to survive in such an instance, and inevitably it will be the higher-cost, more dispersed rural locations that will see the highest rises in prices. In the case study the fare cuts took place primarily in the most populous areas. For wealthier rural residents, this may be a price they are prepared to pay for living in their chosen location, perhaps offset by lower financial levies of whatever sort from local government; for the poor, this will represent a further major burden, and worsen deprivation and social polarization.

Denationalization is perhaps rather less significant as a unique rural factor, although private companies with fewer voluntary or imposed social responsibilities may be likely to take a more hard-nosed view of rural provision. In some sectors, though, there could be particular impacts on landscape (for example, the privatization of the Forestry Commission) or wildlife conservation. The main lesson to be drawn is that, whatever its effects in terms of efficiency (and as has been shown, these are much disputed), an unregulated free market is likely to have severe impacts on the poor and powerless, whether in rural or urban areas. Judging by the transport example, few benefits appeared to reach the actual customers of the industry.

The question then turns, finally, to one of the degree to which these free markets are to be regulated in the future, what form of regulation will be applied and whether a planning role remains of relevance in maintaining and widening rural opportunities. In most cases, some degree of regulation has remained or been devised, and it is vital that such roles and agencies are themselves active and vigilant, and are allowed the powers and resources to perform their stated duties effectively. Imposing controls of any significant kind is anathema to the free-market liberal wing of the New Right, while the controls the moral conservatives might desire are likely to be authoritarian rather than progressive in intent. One of the connections that the privatization initiative undoubtedly made successfully was with a feeling of alienation and powerlessness among customers of the nationalized industries and local government, and indeed a critique of this from a left-wing political perspective, already visible, has been given a strong incentive by the attempt to develop a powerful, progressive alternative to the policies of the current government. Despite the empowering potential of some initiatives, power wrenched from the formerly nationalized firms and local councils has tended to pass to central government rather than to the consumer; indeed, it has been claimed that the least developed forms of privatization are those that would empower consumers against producers most radically (Saunders and Harris, 1990). This provides a positive hook on which to develop alternative thinking, but also implies continual public pressure to keep progressive regulations properly operative.

The planning system also has some potential for redressing rural imbalances, and it has been demonstrated that the role of the county councils is crucial in making the new transport framework operate satisfactorily. Some in particular have been highly innovative within their constraints, and have allowed an ideology to be implemented on the ground without a major initial catastrophe. Planning could have a similar function for other services, but this has been weakened under the present regime. Proposals for further modifications could worsen this effect, but something of a backlash has developed recently, even among elements of the Conservatives, as the consequences of a free market in planning have become evident in both urban regeneration schemes like London Docklands, and the pressured rural growth areas such as central Berkshire. There are also faint glimmers of a new approach on matters such as rural social housing. Like many matters related to privatization and deregulation, these are up for debate, however, and the direction these trends will take is not yet entirely clear.

Conclusion

This chapter has indicated the character of, and justifications advanced for, that complex and sometimes contradictory phenomenon that carries the label of privatization. It has also attempted to evaluate what might happen as this broad theory and approach is implemented in practice in rural Britain. The reality has proved to be rather less dramatic than extremes of opinion might suggest, but the longer-term trends revealed are certainly of concern to anyone who aspires to social equality in the countryside, and indeed elsewhere. Whilst the overall approach might be changed, there are potential mechanisms for counteracting adverse effects instead of (or until) this happens. The effectiveness of such mechanisms will be absolutely crucial for rural areas, and provides a priority area for research attention as successive privatization initiatives work through to affect particular places and people.

10

POPULATION AND SOCIAL CONDITIONS IN REMOTE AREAS: THE CHANGING CHARACTER OF THE SCOTTISH HIGHLANDS AND ISLANDS
Alison McCleery

One of the remarkable features of the last quarter of a century, as noted in Chapter 1, is that rural population and employment growth has not been confined to the countryside around large cities but has penetrated into remoter locations, including of the most peripheral parts of national territories. This process was first observed in the USA, where Morrison and Wheeler (1976) related the rural revival to the attraction of retirement migration, the development of tourism and recreation, the strengthening of national-defence facilities and the exploitation of new fuel sources at a time of high energy prices. Such factors have also been found to have played a significant role in many similarly remote regions in other parts of the Developed World, where during the 1960s and 1970s a long history of economic decline and depopulation seemed to be going into reverse. Subsequent events, however, have cast doubt on the permanence of this revival, with traditional problems reasserting themselves and with the revival itself now being seen as a mixed blessing.

This chapter documents this type of experience for one classic remoter rural region – the Highlands and Islands of Scotland. It demonstrates clearly the geographical peripherality and economic marginality of this part of Britain, but points out that this situation is by no means unique, least of all around the North Atlantic fringe. It shows how the economic developments of the 1960s and 1970s – including the establishment of manufacturing industry and nuclear research and development, the encouragement of tourism and other service industries and, most notably, the arrival of North Sea oil- and gas-related activity, together with the attendant social changes such as age-selective migration and structural ageing – altered the trend and geographical pattern of population and prosperity in this marginal region. The chapter goes on to review this experience in the light of the rather different context of the 1980s and describes the continuing 'quality of life' problems faced by people in most parts of the Highlands and Islands. The chapter begins, however, by outlining the nature of the 'remote area' problem and the policy challenge it represents.

Tackling the problems of remote areas

Solutions to the problem of under-development in the context of an advanced economy may vary according to the perception of the problem and the politics of those doing the perceiving. Often there is a fundamental divergence between how the situation is interpreted at the metropolitan centre on the one hand and by the local community on the other. In general, domination by the metropolitan core is evident not only in terms of the manifestation of the problem of marginality but also in the framing of the solution to that problem, since the periphery lacks not only economic but also political clout. Policies emanating from the centre designed to eradicate marginality inevitably tend to subscribe to the thesis of economic dualism, which holds that marginal regions remain poor because their semi-traditional economies are isolated from the market forces that regulate the modern economic sectors of their constituent societies. Solutions therefore aim to modernize the backward economy and integrate it with the core. However, critics of this approach contend that far from being isolated, marginal regions are already incorporated within the metropolitan economy on terms of severe disadvantage. It is argued that, until it is understood by the decision-makers at the centre that the development of a marginal area involves changing its structure of integration with the dominant economic sector, then regional development policy will simply reinforce pre-existing structures of economic dependency (Prattis, 1977). While policies remain only partially successful for whatever reasons, and as long as peripheral regions continue to manifest economic problems, the arguments as to the precise nature of the relationship of the marginal region with the core will persist.

That there is no consensus on the matter of the 'correct' structural framework for the alleviation of the symptoms of marginality does not invalidate the concept of marginality itself, and it is certainly possible to identify a number of defining characteristics of marginal areas – geographical, economic and social hallmarks about which there is no dispute. The most obvious of these characteristics is physical remoteness that, combined with difficult topography, produces geographical isolation. The coincidence of drowned coastlines with mountainous terrain gives rise to the rias of Brittany, the sea-lochs of the Highlands and the Norwegian fjords, and situations in which some communities (although visible to each other) are nevertheless uncompromisingly severed, remote from each other and isolated in terms of overland transport. In the case of the many actual island locations, accessibility is even more compromised.

The inevitable concomitant of remoteness is economic isolation engendered by distance from product markets, which puts upward pressure upon marginal costs. The double handicap is faced of imported raw materials incurring a transport premium and of higher transport costs for exported finished goods, with the result that potentially lower labour costs are effectively cancelled out. Hence the search for resource-based industries and/or high value-added, low-bulk products. The problems of geographical remoteness and economic isolation are made worse in turn by both climatic excesses and infrastructural deficiencies. On the one hand, cold, wet and unpredictably stormy weather not only shortens the growing season and damages crops but also curtails many economic activities. On the other hand, investment in transport, shopping and social facilities is difficult to justify – and is therefore comparatively absent – in the case of a small global population arranged in widely scattered, low-density communities.

Constrained by populations that are both numerically low and highly dispersed, remote rural areas invariably also demonstrate unfavourable population structures. In that it at once results from and contributes to marginality, the population condition of such an area is not only especially sensitive as an indicator of socioeconomic health but it is also particularly complex to interpret. In the first instance, the typically limited natural resources of a remote rural area characterized by difficult terrain and severe climate are likely to support low numbers of people in primary-sector activities – agriculture, forestry, fishing and mining – which offer limited and fluctuating financial returns. The device of occupational pluralism is regularly needed to augment the inadequate living from subsistence farming. Despite endemic under-employment and seasonal unemployment, opportunities for secondary activities that add value, such as processing, or tertiary activities offering enhanced returns are severely limited or non-existent. With the population thus pressing upon the means of subsistence, and few or no possibilities of employment, let alone better opportunities leading to an improved standard of living, an outflow of school-leavers to seek their fortunes in the metropolitan centres of the core is inevitable. This brings about total population decline on the one hand and structural ageing on the other. As the dwindling population ages, fertility is reduced so that the generation of outmigrants is not replaced. The process gains momentum and eventually a situation is reached where community viability is threatened by the lack of people falling within the economically active/reproductive age cohorts.

As the older generation fades away, with them may disappear a unique linguistic and cultural identity. This sharply contoured cultural specificity is the antithesis of universally flat metropolitan attitudes and lifestyles and is a final distinguishing feature of the marginal region, born of generations of people carving out their existence in isolation in the periphery, separated from the core by the natural barriers of distance, mountains, rivers and lochs. The indigenous cultures of remote rural areas are under threat, even where population is not already in decline. Because modern channels of communication are highly developed, metropolitan mores can and do penetrate into the farthest reaches of geographical remoteness, by means of television, tabloids and tourists. The expectations and aspirations of the inhabitants of remote rural areas are inevitably altered. Second-home ownership is doubly damaging in that it operates on two levels, driving out the indigenous population in the first place by blocking housing opportunities and thereafter slowly eroding the weakened remnants of indigenous culture. Evaluated by metropolitan criteria, the life chances the periphery can offer by comparison with the core are increasingly interpreted as being deficient. Such an unfavourable comparison often provides the critical pull factor that, combined with existing economic and social pressures already acting to push people out of rural areas, strengthens the motivation for migration and generates the momentum for inexorable rural exodus (Byron, 1988).

The example of the Highlands and Islands

It would be difficult to find a better example of a remote rural area than the Highlands and Islands. It is situated on the north-west periphery of Scotland, which in turn forms the northern extremity of Britain, in its turn an island off the western edge of the European continental landmass (Figure 10.1). Yet as an example of a region that is peripheral in terms of geographical location and marginal in terms of economic viability, the Highlands and Islands is not unique and should be viewed within the wider framework of a continuum of such North Atlantic marginal regions, encompassing not only the whole of the Atlantic periphery of north-west Europe but taking in also the Canadian Maritime provinces. Places fringing the Atlantic margins as superficially diverse as mid-Wales, the West of Ireland, North Norway and Nova Scotia all share with the Scottish Highlands the label of being backward areas in advanced countries as well as the common inheritance of a plethora of official policies designed over the years to tackle that perceived problem.

At the same time, the Highlands and Islands could claim to be the oldest development area in the UK (Turnock, 1969). As early as 1746 a treatise on the state of the Highlands highlighted the generally prevailing circumstances of primitive hardship, while Pennant (1774, Vol. 2, p. 228) described levels of penury and squalor that could fairly be compared with those of a present-day Third World famine area – conditions that were tolerable only because they were traditional and familiar (Gaskell, 1968). Following the devastation caused by the collapse of the kelp industry, the failure of potato crops and the 'clearances' of the nineteenth century, the Congested Districts Board (CDB) for Scotland was set up in 1897, modelled on the system used earlier in western Ireland. Charged with developing agriculture and fishing, creating holdings for landless cottars and enlarging existing holdings, improving the communications infrastructure and fostering home industries, the CDB was the first of many attempts to stimulate Highland development; Highlands and Islands Enterprise is but the latest. In between there has been a series of proposals, a few of which have been realized as policy, notably the Highlands and Islands Development Board (HIDB) (for further details, see Campbell, 1920; Gray, 1957; Magnusson, 1968; Mackay, 1973; Williams, 1973; McCleery, 1984).

From population decline to revival[1]

When the HIDB was established in the mid-1960s, the Highlands and Islands were exhibiting the effects of a century of population decline, reflecting the area's long-term economic problems. By 1961 the population had fallen by some

30 per cent to 302,000, from a peak of 424,000 in 1851. This contrasts with an increase of over 70 per cent in the population of Scotland as a whole. Between 1921 and 1961 alone the Highlands and Islands lost almost a quarter of its population. The concomitant of heavy outmigration was increasing age structure imbalance. Long-term population decline may also have led to some weakness in community leadership and a loss of entrepreneurial talent. The rate of population loss reflected the deep-seated problems of the regional economy of the Highlands and Islands at a time of full employment and sustained growth in the national economy. Unemployment in the Highlands and Islands in the 1950s and 1960s was about twice as high as in Scotland as a whole, and four times as high as in Great Britain. Employment was heavily dependent on the primary sector, which manifested both below-average incomes and decreasing opportunities for employment. Under-employment was also a problem, remoter areas exemplifying occupational pluralism as a device to maintain an adequate level of income. In the absence of migration as a safety valve, levels of unemployment in the Highlands and Islands would probably have been considerably higher.

Thus were perceived the major difficulties of the area at the time when the decision was taken to establish the HIDB. However, it is pertinent to observe that the long-term decline of the area's population was broadly arrested, certainly at aggregate level, before either the board's activities or North Sea oil developments became major factors. Earlier attraction of major industrial projects – notably the United Kingdom Atomic Energy Authority research facility at Dounreay and the integrated pulp-and-paper mill at Fort William – together with improved communications and the growth of tourism probably contributed to this change. In consequence, the 1960s was characterized by a redistribution of population within the area, with the Moray Firth and Lochaber gaining at the expense of other parts, although the actual migration pattern was not so straightforward. Thus the growth of Inverness as a regional centre and of Easter Ross as an industrial centre was only weakly related to continued marked decline in the population of the Clyde Islands and in the remote north and west fringes of the area. For example, there was still in the 1960s substantial outmigration from the Highlands and Islands and, if net population decline over the 1961–71 period was only marginal, this is because natural increase largely compensated for the loss. Already by 1971 there was evidence of slight recovery from the low of around 300,000 indicated by the 1966 Sample Census.

It was during the 1970s, however, that there was a dramatic turnaround in population trends in the Highlands and Islands. Between 1971 and 1981 the population increased on the most conservative estimate by some 25,000 (8.4 per cent), or on the most generous by some 38,000 (12.5 per cent) – almost wholly as a result of substantial net inmigration. Whichever 1981 population base is employed, natural change accounts for an increase of less than half a per cent.[2] The bulk of the growth took place in three areas, Easter Ross (32.5 per cent), Shetland (29.4 per cent) and Inverness (12.1 per cent) – whose combined population rose by almost 21,000. Wester Ross, too, experienced rapid growth (23.2 per cent) but, because of a much smaller base population, in absolute terms this represented fewer than a thousand people. These were the areas to benefit directly and indirectly from oil-related development.

However, the 1970s turnaround was not confined to the areas of oil impact and included a wide range of localities with little or no oil-related development, although growth in areas such as Badenoch and Strathspey (7.9 per cent), Oban and Lorn (7.6 per cent) and even Arran and the Cumbraes (15.9 per cent) was much more modest (Figure 10.1). As well as increasing during the 1970s, the population structure of the Highlands and Islands also aged. The overall increase of 8.4 per cent incorporated an 18-per-cent increase in the numbers of very elderly (75 years and over) and a corresponding decline of 8 per cent in the very young (under 5 years). However, the fact of an ageing population structure, in common with most Western populations during this period, should be set against a very healthy 23 per cent increase in 25–44-year-olds, the key working age cohort. As Table 10.1 shows, relative to Scotland as a whole, age structure in the Highlands and Islands has improved markedly between 1961 and 1981.

Economic structure and employment

Reflecting the transformation in the Highlands and Islands economy, the employment structure of the Highlands and Islands has moved some way towards the national average but still remains heavily weighted in favour of the primary sector. Agriculture, fishing and tourism, practised either separately or as the form of occupational pluralism better known as crofting, are the mainstays of the economy of the remoter north-west mainland and islands and self-employment therefore features prominently. The service and construction sectors of rural economies also tend to be relatively large and in this the Highlands is no exception. Indeed, in 1981 services accounted for two thirds of total employment in the area, higher than for Scotland as a whole and much higher than for other Scottish rural areas (Table 10.2). This proportion has risen steadily over the longer term, with an especially sharp, oil-induced rise in the 1970s. The sector primarily serves the needs of the local population but there is also an important 'tradable' element in the form of the tourist industry. Tourism

Table 10.1   Changes in age structure in the Highlands and Islands* relative to Scotland (%)

| | 1961 | | 1971 | | 1981 | |
	Scotland	H & I	Scotland	H & I	Scotland	H & I
Ages 0–4	9.0	8.1	8.5	8.0	6.1	6.8
Ages 25–44	25.6	23.7	23.5	22.7	25.6	25.7
Ages 75+	3.7	5.6	4.2	5.7	5.3	6.2

Note * HIDB area pre-1986.

(*Source: Census of Population.*)

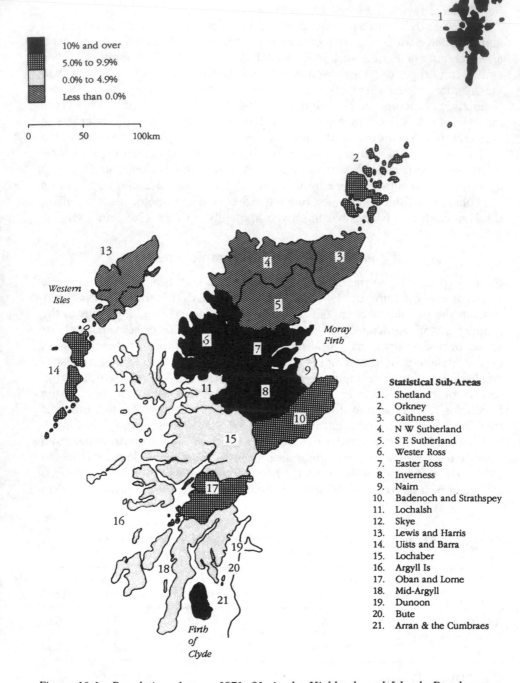

Statistical Sub-Areas

1. Shetland
2. Orkney
3. Caithness
4. N W Sutherland
5. S E Sutherland
6. Wester Ross
7. Easter Ross
8. Inverness
9. Nairn
10. Badenoch and Strathspey
11. Lochalsh
12. Skye
13. Lewis and Harris
14. Uists and Barra
15. Lochaber
16. Argyll Is
17. Oban and Lorne
18. Mid-Argyll
19. Dunoon
20. Bute
21. Arran & the Cumbraes

10% and over
5.0% to 9.9%
0.0% to 4.9%
Less than 0.0%

0   50   100km

*Figure 10.1   Population change, 1971–81, in the Highlands and Islands Development
Board Area (pre-1986), by Statistical Area.*
Note: Data refers to usual residents. Overall rate for HIDB Area is 8.4 per
cent.

Table 10.2 Employment structure in the Highlands and Islands* relative to Scotland and other rural areas, 1981

	Highlands & Islands (No.)	Highlands & Islands (%)	Scotland (%)	Borders (%)	Dumfries & Galloway (%)
Primary	10,950	8.9	4.6	10.8	11.3
Manufacturing	16,760	13.6	24.8	31.1	22.9
Construction	12,710	10.3	7.4	5.7	5.7
Public utilities	1,630	1.3	1.5	1.1	2.7
Services	81,000	65.8	61.6	51.4	57.4
Total	123,060	100.0	100.0	100.0	100.0

Notes * HIDB area pre-1986.
Figures relate to employees in employment except in the case of fishing where Department of Agriculture and Fisheries for Scotland (DAFS) estimates of self-employed are included.

(*Source: Census of Employment*; DAFS.)

is an important source of employment in the area (with an estimated 12,500 direct jobs excluding the self-employed and proprietors), although the season is short and influenced by the weather. Employment in other services – consumer services, business services, education, health and administration – is largely dependent on population trends and slower growth therefore seems likely in future.

Activity in the construction industry has fallen of late, in part because of the completion of oil-related projects, but the sector remains an important employer. This reflects the problems and costs associated with construction projects in remote areas as well as the investment in the region's infrastructure, which has taken place over the years. Finally, the manufacturing sector in the Highlands and Islands is relatively small and widely scattered. In 1981 it accounted for some 13 per cent of the region's total employees in employment, compared with a Scottish average of 25 per cent, and 31 per cent in the Borders region. Distance from markets and sources of supplies and services, limited labour markets and transport and communications problems all militate against the growth of manufacturing industry, which tends to be characterized by a large number of very small units at one end of the scale and, at the other, a few large units, as, for example, round the Moray Firth. In 1983, 69 per cent of manufacturing units in the Highlands and Islands employed fewer than ten people as compared with the Scottish average of 58 per cent. Businesses in this category tend to be based on the development of local resources, embracing food processing, textiles, whisky, timber and a small engineering and oil-related sector.

Three distinct phases may be identified in the post-war history of employment in the area. Growth during the 1950s and 1960s was in line with Scotland although below Great Britain. Major projects at this time included the Dounreay nuclear facility, the pulp-and-paper mill at Fort William and the Invergordon aluminium smelter. A dramatic growth of 25 per cent in the area's employment during the 1970s was largely attributable to oil-related development, the main beneficiaries being the eastern seaboard and the Northern Isles, and to a lesser extent Wester Ross and the Western Isles, a situation reflected in the pattern and process of population redistribution as discussed above. The peak in oil-related employment in the area was reached in 1981 and tailed off sharply thereafter with the completion of the oil terminals at Sullom Voe (Shetland) and Flotta (Orkney). The 1986 oil-price fall occasioned further sharp reductions, so that by the start of 1987 oil-related jobs were down to as few as a quarter of the 1981 total of 16,500.

The volatile post-war employment situation is reflected in the changing unemployment levels during the same period. During much of the 1950s and 1960s, as already mentioned, unemployment in the Highlands and Islands was roughly twice as high as in Scotland as a whole and four times the Great Britain rate. The gap in unemployment rates began to narrow from the later 1960s and closed dramatically during the 1970s as unemployment nationally increased more rapidly than in the Highlands and Islands. Clearly, oil-related development had a major impact on these trends and by 1981 the unemployment rate in the area was significantly below that for Scotland. During the 1980s the trend in unemployment in the Highlands and Islands has been less favourable, the rate tending to rise more rapidly and latterly fall more slowly than for Scotland as a whole (Table 10.3). As a result, its unemployment rate relative to the Scotland rate (Scotland = 100) deteriorated from 84 in April 1981 to 103 in April 1988. In other words by 1988 the unemployment rate in the Highlands and Islands, though by now beginning to fall in absolute terms, had overtaken the Scottish rate, which itself exceeded those for Great Britain and for comparable rural areas such as the Borders or Dumfries & Galloway. However, by April 1989 the unemployment relative was back to 96, reflecting a dramatic 3.7 per cent annual fall in unemployment in the Highlands and Islands as the upturn in employment

Table 10.3 Unemployment in the HIDB area compared with Scotland and Great Britain, 1981–90 (April) (%)

	HIDB	Scotland	Great Britain
1981	10.8	12.8	10.3
1986	15.3	15.8	13.6
1987	15.4	16.4	12.5
1988	14.4	14.0	10.1
1989	10.7	11.2	7.4
1990	9.0	9.4	6.4

Note The HIDB area was extended with effect from November 1986 to include 13 parishes in Moray district.

(*Source:* HIDB for HIDB data; DE for Scotland and Great Britain data.)

there suddenly caught up. Within the Highlands and Islands itself, rates can vary substantially, reflecting the considerable differences in employment structure. For example, at March 1987 more than one in four people in Skye and Wester Ross were seeking work as compared with less than one in fourteen in Shetland.

Population trends in the 1980s

Analysis of population change post-1981 is necessarily based upon the Registrar General's mid-year estimates, which are not disaggregated below local-authority district level. Furthermore, the annual estimates adopt a different convention from the Census in defining the usual address of students, members of the armed forces and, most importantly in the case of the Highlands and Islands, workers in construction camps. It is estimated that during the period 1981–5 alone, Shetland lost some 3,000 temporary construction workers, accounting for virtually all of the 11 per cent net decline there during these four years. The Shetland experience inevitably dominates the broader picture, depressing the overall change figure for the 1981–9 period to +1.0. If Shetland is excluded from an overall change calculation, the rate of population growth increases to 2.4 per cent (annual average 0.3) as compared with an all-Scotland change figure for the same period of −1.7 per cent (annual average −0.2). However, as Table 10.4 indicates, with the clear exception of Inverness, none of the individual

Table 10.4   Post-1981 population of the Highlands and Islands by local-authority district*

	1981	% change	1985	% change	1989	1981–9 % change
Shetland	26,347	−11.0	23,440	−5.4	22,170	−15.8
Orkney	19,182	0.9	19,351	0.2	19,400	1.1
Caithness	27,636	−1.2	27,302	−2.7	26,560	−3.9
Sutherland	13,313	−0.6	13,238	−2.1	12,960	−2.6
Ross & Cromarty	46,924	2.1	47,889	0.9	48,310	3.0
Inverness	57,105	3.1	58,849	5.8	62,250	9.0
Nairn	9,953	2.3	10,180	0.6	10,240	2.9
Badenoch & Strathspey	9,860	5.9	10,442	4.2	10,880	10.3
Skye & Lochalsh	10,621	6.5	11,308	2.5	11,590	9.1
Western Isles	31,548	0.0	31,545	−2.9	30,630	−2.9
Lochaber	19,491	−0.4	19,409	−1.5	19,110	−2.0
Argyll & Bute	65,018	1.3	65,873	0.7	66,360	2.1
Arran & Cumbraes[†]	5,100	0.7	5,136	−2.1	5,029	−1.4
Highlands & Islands	342,098	0.5	343,962	0.4	345,489	+1.0
Scotland	5,180,200	−0.8	5,136,509	−0.9	5,090,700	−1.7

Notes * This area plus Arran and the Cumbraes is the same as the pre-1986 HIDB area.
[†] Estimates provided by Cunninghame District Council.

(*Source:* Registrar General's mid-year estimates of population.)

authorities to demonstrate population increase during the 1981–5 period was able to better its performance in the 1985–9 period. Similarly, with the clear exception of Shetland, all of the authorities to demonstrate population decline during the earlier period showed even faster decline in the later period. Furthermore, the annual estimates series shows that the overall Highlands and Islands population actually peaked in 1986 at a figure of 345,717, thereafter levelling off. To sum up, population growth during the 1980s had clearly moderated, yet was still in evidence, in contrast to the overall Scotland position of decline.

The geographical pattern of Highland population growth has also altered considerably since the 1970s. The dramatic 1970s gainers of Ross & Cromarty and Shetland have ceded their transient supremacy. Skye & Lochalsh, Badenoch & Strathspey and especially Inverness continue to prosper, the latter now firmly established as the region's industrial, commercial and retail service centre, and Nairn, another eastern seaboard urban authority previously supported by a buoyant oil-based economy, is joined by Ross & Cromarty in achieving more restrained, yet still recognizable, growth. Although at the top new leaders have emerged during the 1980s, unfortunately at the bottom the familiar geographical bias in demographic fortunes is still evident. Shetland excepted, all of those areas to perform less well than Scotland – namely, Caithness, Sutherland, the Western Isles, Lochaber and Arran & Cumbraes (post-1985 only) – come into the category of that most vulnerable peripheral and insular north and west fringe. Lochaber, with the lowest percentage decrease, may in fact have the most serious problem, given that this is an example of already sharply reduced growth during the 1971–81 period being transformed into outright decline, albeit modest.

Compared with the previous decade, therefore, the 1980s are seen to be characterized by a narrowing of the range of areal variation in population change within the Highlands and Islands. The area-specific dramatic growth of the late 1970s associated with oil was over, and growth rates generally were more restrained, these two strands of experience being consistent on the one hand with the direct effects of the ending of the construction phase of oil exploitation and, on the other hand, with the indirect reverse multiplier effects of this same run-down. Finally, a situation of aggregate population growth during the 1980s in the Highlands and Islands should not be allowed to mask the deceleration in the rate of that growth to the extent that, by the close of the decade, the population was static. Nor should it detract from the seriousness of the position in some parts of the remote north and west Highlands, which are now struggling to maintain their previous levels of population and exhibit age-structure problems that underline their essential demographic fragility.

A further indication of this demographic fragility is provided by the age structure of migration, as revealed by the National Health Service Central Register (NHSCR) data on flows between Health Board Areas. Table 10.5 shows annual ratios of inmigrants to outmigrants for all ages and for selected sub-groups for the Highland region. In terms of all ages, it is found that the Highland region witnessed net inmigration throughout the decade, with the ratio

Table 10.5   Ratio of inmigrants to outmigrants for Highland Health Board Area*

	All ages	0–14 yrs	20–24 yrs	60–74 yrs
1979–80	1.01	0.98	1.08	1.63
1980–1	1.13	1.11	1.16	1.72
1981–2	1.07	0.98	1.23	1.40
1982–3	1.18	1.27	1.20	1.65
1983–4	1.17	1.11	1.26	1.84
1984–5	1.22	1.16	1.22	2.16
1985–6	1.19	1.24	1.13	1.39
1986–7	1.09	1.07	1.10	1.92
1987–8	1.03	0.97	0.98	1.91

Note * Highland HBA is consistent with Highland region and includes the following eight local-authority districts: Caithness, Sutherland, Ross & Cromarty, Inverness, Nairn, Badenoch & Strathspey, Skye & Lochalsh and Lochaber.

(*Source:* Derived from NHSCR data, kindly supplied by General Register Office (Scotland).)

peaking at 1.22 in mid-decade, thereafter falling back to barely above unity. However, the key migration group in terms of volumes of gross flows is the 20–24 years cohort, making up about 17 per cent of total movement. The ratio for this group also peaked in mid-decade but at 1.26, subsequently falling back to slightly below unity. That is, not every young adult leaving was able to be replaced. In other words, overall change through migration is reinforced by a structural effect, net gain being disproportionately associated with the gain of younger people and vice versa, just because young people make up the bulk of migrant flows. The pattern for the ratio for 0–14-year-olds also parallels the overall pattern but is fairly consistently less healthy, both beginning and ending the series below unity. By contrast, the ratio for the 60–74 age group is much higher and has tended to rise throughout the decade, varying between 1.4 and 2.2. That is, by the late 1980s there were twice as many of this group of elderly moving in as moving out, which, in the context of overall net loss, is especially unsatisfactory – even if absolute numbers are small. Jones *et al.* (1986) produced conclusive evidence to dispel the myth that the elderly figure prominently in long-distance migration to the Highlands. Nevertheless, it is the combination of a longstanding annual excess of elderly migrants with volatility in respect of key youth cohorts that poses the threat to a population already ageing as a result of natural change.

The aggregate pattern which emerges from these analyses is very complex, consisting of many migration sub-systems often linked to highly specialized labour markets. Shetland constitutes the most pronounced but by no means the only example of this phenomenon, since Caithness and Lochaber share aspects of the Shetland experience. The eastern seaboard authorities, possessing many of the characteristics of the urbanized, industrial economy, form another distinctive grouping, as also do the remote western and northern areas, with their crofting traditions and bias towards primary and self-employment. Finally, there are the Central Highlands and Argyll, the economies of which are largely based upon hill farming and the tourist industry. To summarize at the risk of over-simplification, within the area there is a continuing trend of population redistribution, with losses from remoter parts in favour of the Moray Firth growth centres. Until 1981 this drain was offset at least in part by inmigration from elsewhere in Scotland, as well as from the rest of Great Britain, but there are indications that this pattern has not persisted throughout the 1980s. Particularly for remote areas such as Lewis and Harris, the decade ended on a note of uncertainty.

Quality of life

Quality of life is an elusive concept that relates to something transcending the material concerns of everyday life (Pacione, 1984b). It was during the 1970s that mounting dissatisfaction with traditional measures of economic health, such as rates of unemployment and outmigration being used as surrogates for a general measure of social well-being, led to the emergence of a new field of research into social indicators. Yet, although there is now a quite considerable body of literature on social indicators, none of it has revealed any simple way of expressing the state of well-being of a given area (Cottam and Knox, 1982). Even where a social-accounting approach to regional development is adopted, rural areas are disadvantages by the dominance of urban concerns in influencing social policy. Early in the 1980s, therefore, the HIDB commissioned a study to address the problem of developing a series of quality-of-life indicators for their area.

The resultant report by Cottam and Knox (*ibid*. p. 39) emphasizes that

> the increasing vulnerability of parts of the Highlands and Islands to adverse changes in social well-being may be traced in the main to a contraction in population over many years precipitating the withdrawal of services and facilities and, much less frequently, to a sudden upsurge in economic fortune attended by high levels of in-migration and social upheaval.

Population decline is seen to be responsible for initiating a downward spiral of deprivation, which traditional regional policy has been powerless to arrest. The associative relationship between declining population and declining opportunity went unrecognized for decades, depopulation being attributed almost solely to a shortage of land and to a lack of employment but rarely equated with perceived opportunity or quality of life. For example, people living in deprived rural conditions are less inclined to undergo training or seek jobs – a situation aggravated by the tendency for successful families to leave. Opportunities for individual improvement are thus further reduced, with consequent de-skilling and reduction in the (already weak) prerequisites for economic development.

In setting out to devise a means of quantifying this process, Cottam and Knox looked initially to Shaw's 1979 threefold categorization of rural deprivation. His three dimensions of household deprivation, opportunity deprivation and compromised accessibility do not pretend to be either comprehensive or mutually exclusive. Thus, while the structural quality of housing is important – a fact substantiated by Cottam and Knox's own inclusion of sub-tolerable housing in their primary-diagnostic indicator group – so is the educational experience or other personal circumstances of householders, such as age or income, which can affect their ability to take advantage of available opportunities. In 1983 the Western Isles Islands Authority estimated that over 50 per cent of private houses in the Uists and almost 70 per cent in Barra were below tolerable standard (Shucksmith, 1988). Particularly in the crofting areas, high levels of owner-occupation associated with substantial housing-amenity deficiency do reflect the fact that many owners are either elderly or on low incomes. For example, in 1981 9 of the 21 Highland Statistical Areas had more than one in five pensioners – a level comparable with that in the retirement areas of the south coast of England.

The link between compromised accessibility and opportunity deprivation needs little amplification. Physical remoteness equates with social isolation, since it reduces the possibilities for communication and information dissemination, leading in turn to an inability to capitalize upon available opportunities – a criticism often levelled at the indigenous inhabitants of rural areas whose perceived inactivity seems to contrast directly with the initiative and drive of incomers. In reality, extremely uncomfortable choices are faced by households with low incomes in remote rural areas, which are avoided even by their urban counterparts. Those households without private transport are severely disadvantaged in respect of access to distant service centres, while for some households with private transport, the outlay on purchase, maintenance and running costs may well prejudice other aspects of quality of life. In other words, in respect of access to facilities there is a very real opportunity cost to be borne by the inhabitants of rural areas.

Stated simply, the paradox of life in the HIDB area is that the cost of living is generally higher than elsewhere (7 per cent greater than in Aberdeen or Edinburgh in 1986) in an area where per-capita income is generally lower than elsewhere (95 per cent of the Great Britain rate for males at 1985, 90 per cent for females). Furthermore, because of the structure and sparsity of population, the need for services is actually greater, yet the possibility of provision less. Particular difficulties exist in access to health and education services. Hospital treatment in urgent and non-routine cases may require travel to Inverness, Aberdeen or Glasgow, while in the case of education, weekly boarding provision is necessary in a number of areas for pupils from remote locations who are unable to travel on a daily basis to school. In other words, dislocation in space of demand relative to supply constitutes the essential problem of quality of life in much of the Highlands. This is the crux of what, albeit construed primarily in demographic terms, has earned for itself the sobriquet of the 'Highland Problem'.

Conclusion

The foregoing analysis of population and related change in the Scottish Highlands and Islands has outlined the improvements that have taken place there over the past quarter century or so. However, it is clear that the sharp population and employment increases of the 1970s were the results of special factors unlikely to be repeated. Given the recent downturn in the oil-related sector in particular, the general economic prospect for the area would seem to be for some weakening in relative performance compared with the immediate past. Some of the former growth points of the 1960s and 1970s including Inverness, which as regional service centre must now sustain itself in the context of erosion of its industrial base, may expect to experience problems as they adjust to economic change. Equally, the northern and western periphery seems likely to continue to suffer from outmigration as a result of the fragility of the economic base of remote rural and island communities. Increasingly, there are only slender grounds for optimism that population in the area as a whole will remain reasonably stable over the medium term, experience suggesting that the development of local resource-based industries would seem to offer the best prospects for future employment opportunities.

It is valuable to view the situation of the Highlands and Islands of Scotland within the framework of the experience of North Atlantic marginal regions in general. The position is most neatly summarized by O'Cinneide (1988, p. 97), who pulls no punches in stating that

> The consolidation of the periphery during the 1970s has not continued through the 1980s and the so-called turnaround trend has been so dramatically reversed that it is now regarded as only a temporary interruption of the long term decline of marginal regions. Adverse demographic trends such as increased out-migration, reduced in-migration, ageing and decline of overall population again are dominant in the periphery. The policy goals of reducing disparities between core and periphery thus stabilizing the rural settlement pattern now appear more elusive than ever.

If official development policies in marginal regions have not been very effective, he continues, it is because they tend to represent inappropriate impositions on these regions by policy makers who are not *au fait* with their real needs and potentials. They have failed to involve the local populations in the development process and, as such, have contributed to the development of a dependency mentality in these areas. Instead, a bottom-up approach to the development of remote rural regions is advocated, which will involve the indigenous population in the identification of needs and resources and in the formulation and implementation of development strategies.

In fact the socialist version of this approach has already been tried by the HIDB in the form of the community co-operation scheme implemented during the chairmanship of erstwhile academic economist, Professor Sir Kenneth Alexander. The capitalist version, aimed at stimulating the private sector, has its own opportunity with the inauguration of Highlands and Islands Enterprise in April 1991. This umbrella organization will continue to be chaired by Sir Robert Cowan and to provide a variety of central services, but the development function of the former HIDB will now be assumed by a network of some eight local enterprise companies led by directors, managers and owners of, or partners in, locally based businesses. The first decade of the HIDB's life was dominated by growth poles and manufacturing industry, which epitomized the 'white heat of technology' thinking of the time, while the second was ushered in to plans for resource-based industry and the satisfaction of the social and cultural needs of local communities – a concept that was no less a child of its time. Yet without the intervention of oil, the HIDB might well have been condemned to a rather straight and narrow path to somewhat uninspired regional development, with inevitably short-term results. The oil era not only forced a change in direction but also permitted greater freedom to experiment with novel, and potentially risky, strategies – in particular, for revitalization of the remoter periphery.

In shouldering the responsibility for restoring a post-oil equilibrium, Highlands and Islands Enterprise benefits from those two decades of more than usually varied experience. It also starts out from a much higher base than its predecessor did in 1965, since the area is demonstrably in a much stronger economic position than in the 1960s and earlier, in consequence of the substantial investment, both public and private, which has taken place in its general infrastructure and in economic development over the years. Furthermore, the notion of local entrepreneurs being better placed than centralized administrators to assess what is and is not possible in business terms has merit and is perhaps akin to Hansen's (1988) emphasis on the need to see the periphery through the eyes of the native young. But Hansen, equally, calls for a new economic geography of the periphery. 'Just as the emphasis on social processes arose as a reaction against a too simplistic economic geography, my plea for a new economic geography of the periphery is a reaction against a social geography oblivious to the importance of productive forces' (*ibid.* p. 111). In other words, it may be unrealistic to expect all areas in the periphery to be equally successful in the achievement of the desired goal of economic development, not least where, in consequence of the individual decisions of local enterprise companies taken in isolation, they may be competing against each other. As Smout (1972, p. 337) has observed of the Highlands, 'The grim facts of economic geography have, time and time again, defeated the good intentions of planners'. The first chairman of the HIDB once described that institution as a merchant bank with a social purpose. It is the efficient achievement of the correct balance between the two across an extremely large and diverse area that is so elusive.

Notes

1. An expanded version of much of the material presented in this and the following section, including detailed statistical annexes, was originally prepared for the Industry Department for Scotland (McCleery *et al.*, 1987). A summary of key findings is published separately (Walker and McCleery, 1987) as is a context to the study (McCleery, 1988). The author alone is responsible for the views expressed in this chapter, which are not necessarily those of the Industry Department for Scotland.

2. At 1981 neither the usually resident population – the basis for the lower figure – nor the transfer population – the basis for the higher figures – is entirely satisfactory for purposes of comparison with 1971. On the one hand, the 1981 usually resident population – present on Census night or absent from an occupied house – is understated because of the coincidence of Census night with a major holiday period. On the other hand, a change in the 1981 method of allocating wholly absent households to their usual addresses means that there are no comparable 1971 transfer figures. In any case, disaggregated statistics at both dates are available only for the present/absent base so that discussion is, by default, based on the usually resident population.

11

COMMUNITY INVOLVEMENT IN RURAL DEVELOPMENT: THE EXAMPLE OF THE RURAL DEVELOPMENT COMMISSION

Ian Bowler and Gareth Lewis

Since the Second World War, Western rural society has undergone a social and economic transformation commonly referred to as 'development'. The processes of transformation have undoubtedly acted unevenly on rural society (McLaughlin, 1986; Summers, 1986), although considerable ambiguity remains on the exact meaning of the term 'development' (Barsh and Gale, 1982; Welch, 1984). For example, in a wide-ranging review of the concept, Hoggart and Buller (1987, p. 28) eventually concluded that 'the reader must decide whether the trends are developmental or not and in what ways'. However, as Buller and Wright (1990, pp. 1–5) rightly point out, there are a series of common threads that distinguish development, both as a process and as a goal. For example, within broad societal goals development can be 'defined in terms of an overall improvement in the welfare of rural residents and in the contribution which the rural resource base makes more generally to the welfare of the population as a whole' (Hodge, 1986, p. 22). As a process, many observers conclude that any improvement in the welfare of people should also include the means of sustaining it, since 'if people are incapable of sustaining the improvements that have happened, then what occurred was not development but a short-term improvement in living conditions' (Hoggart and Buller, 1987, p. 26).

Recognizing both social and economic welfare dimensions, rural development can be conceived of in two main ways: on the one hand, as a form of structural change and its differential impacts on communities within the countryside; on the other hand, as the role of policies and agencies in effecting social and economic change. In Britain the latter has almost become synonymous with the concept and has tended to dominate both British thinking and research. Indeed, this chapter follows the latter course. It is concerned with recent trends in rural development in Britain, particularly the shift towards greater community involvement with agencies that promote the process. The chapter assesses the effectiveness of this emerging policy emphasis with reference to the attempt by one development agency, the Rural Development Commission, to incorporate such an approach into its work in the rural counties of England.

Policy sectors and agencies

Until recently there has been an overwhelming consensus among politicians and policy makers on the desirability of State intervention as a means of attenuating the effects of uneven economic and social restructuring in the countryside. Generally, this has involved procedures for ameliorating the disadvantages experienced by many rural people, often in defined areas, as well as attempts to reverse economic decline by actively promoting growth in the countryside. Even the recent 'turnaround' in rural population, with its pressure for residential development particularly in the south of England, should not be over-emphasized; in most areas it does little more than 'mask' the issue of rural deprivation (Lewis, 1989).

Within the rural sphere, three alternative models of State intervention for development can be recognized: sectoral, integrated and community-based. In the first model, planned rural development is approached on a sectoral basis by individual government departments, and their dependent agencies, each having a responsibility for a subject area such as forestry, agriculture, transport or tourism. Often the departments and agencies have a regional administrative structure (Yuill, 1982). Over the past forty years, rural policy has been dominated by the agricultural sector, and the resultant changes in agriculture have had widespread effects upon the economic and social well-being of the countryside. A good deal of the remainder of rural policy has been concerned with ameliorating many of the consequences of these changes. Experience has shown that sectoral agencies commonly fail to consult, co-ordinate or collaborate in their operations, even at a regional level; consequently there has been a lack of a common objective and, at times, a duplication of effort and resources (Wright, 1983). It is generally agreed that sector policies have failed to provide across-the-board improvements within the countryside. In such a situation some rural groups have benefited while others have been unaffected; some have even experienced a reduction in their access to employment, housing and services.

In response to the failures of the sectoral approach, an alternative model, integrated rural development, has been adopted (McNab, 1984; Davison, 1985; Rennie, 1986). The concept has been to create institutional structures that bring together the sectoral agencies in a co-ordinated programme of development. During the 1960s, for example, development was seen as a means of overcoming structural weaknesses in capital accumulation; consequently, adjustments in access and resources through the development of growth poles, spread and multiplier effects by regional agencies were the order of the day. Several major rural development agencies were created in the UK at this time, including the Highlands and Islands Development Board, Mid Wales Development, the Council for Small Industries in Rural Areas and, for a short time, the North

Pennine Rural Development Board. These agencies paralleled institutional structures elsewhere in Western Europe (Yuill, 1982). In terms of stimulating factory development and local enterprise, it is often claimed that these agencies have been highly successful. However, it must be admitted that the long-term future of many of these regional initiatives is far from assured, while the agencies' ability to reverse the decline of rural services has yet to be confirmed (Shucksmith and Lloyd, 1982; Williams, 1985). Yet the European Community has embraced the approach by initiating integrated rural development program-mes in areas such as the Western Isles of Scotland (1982), with the aim of bringing together factory building, agricultural policy and infrastructural provision (Clout, 1984).

By the 1980s there was growing scepticism towards many of the activities of these development agencies and a demand began to be expressed for a greater degree of community involvement in rural development. Such dissatisfaction emerged from three sources. First, there was general agreement that those who benefited most from the policies of development agencies were not always those in most need: the prime objective of the agencies appears to have been that of economic growth rather than broader community development. Second, there was a growing disillusionment among local people towards many of the policies of the development agencies (Williams, 1984). According to Wenger (1982), much of this stemmed from a difference between agency and community in the image of developments. In mid-Wales, for example, although there was general agreement that the problem of the region was one of stemming depopulation and increasing job opportunities, views on the means by which this could be achieved differed quite markedly. Mid-Wales Development, working at a regional level, emphasized economic development on the basis of a diffusionist model and a necessary increase in gross national product and per-capita income; on the other hand, the local population conceived development in a more locally based context emphasizing small businesses, the enhancement of local amenities and services and, in large parts of the region, the preservation of the Welsh language. In many ways, 'the type of development planned by the agencies is antithetical to many of the basic values of those who live in the area, who seek to preserve the viability of small communities, ethnic and local integrity, and small scale development' (*ibid.* p. 7). Third, after 1979 there was a fundamental shift in government policy towards less intervention and a greater reliance on the market (Cooke and Hulme, 1988). Inevitably with the decline in the State's interventionist role, and hitherto with only limited private-sector support, development agencies have been forced to emphasize the potentiality of voluntary and small-scale, private-sector action. In other words, from different viewpoints and for different reasons, politicians, planners and local residents have come to recognize the importance of 'bottom-up' or local self-help and community action in the development process. Significantly, this is an international movement and not just confined to the UK (Coffey and Polese, 1985; Ross and Usher, 1986; Douglas, 1987; Bamberger, 1988; Murray and Hart, 1989).

Community development and self-help

A striking feature of the last decade has been the emergence of greater community-based development within the countryside. In this shift to 'localism' and a more 'bottom-up' approach to development, much emphasis is placed on community participation and in the power of local decision-making (Deavers, Hoppe and Ross, 1986). According to Holdcroft and Jones (1982, p. 211), such community development is 'designed to encourage self-help efforts to raise standards of living and create stable self-reliant communities with an assured sense of social and political responsibility'. Grieco (1990, p. 31) goes further by claiming that 'participation must be a key consideration in any attempt to produce genuine development. Simple delegation of responsibility to the existing local power structure may generate as many problems as it resolves'.

Rural community development in Britain has taken several forms, ranging from locally based self-help to more community-oriented policies by the various development agencies (Rogers, 1987). Local self-help, based entirely on the resources of a community, are numerous and diverse. Unfortunately, these initiatives, whether they be a community bus or a community shop, have received considerable media attention and, as a result, there has been a tendency to overlook both their limited scale and reliance on a few volunteers for their operation (McLaughlin, 1987).

In order to initiate greater community involvement and, it is hoped, provide some guidelines for future rural development, two experimental schemes – the Leominster Marches and the Peak Park integrated development projects – were set up in 1981 (Parker, 1984, Leominster Marches Project, 1986). Interestingly, the two schemes are in some ways similar but in others quite different. For example, the management and financial responsibility of the two was alike except that the latter, based on the parishes of Longnor and Monyash, was funded by a group of agencies, including the European Community, whilst the former was sponsored by the district and county councils and the Community Projects Foundation – an independent body funded by the Home Office. Both schemes were based on the premiss that any resource allocation should involve the local communities but, significantly, they differed in the means by which this objective was to be achieved. The Peak Park project appears to have been primarily concerned with developing a more effective institutional framework for the delivery of community-based development. Among its various initiatives were grant-aid packages specially designed to encourage greater community activity, with project leaders appointed to increase community participation, largely by means of exhibitions, public meetings and leaflets. The Leominster Marches project, however, attempted a different approach to development by working 'with' the community rather than 'through' it. The prime objective of this project appears to have been to increase local action in the development process by the enhancement of the residents' ability to identify local needs and to respond to them, particularly their articulation through various bodies offering grants and services. The basis for achieving this objective was an enhancement of local organizational awareness and communication skills – a task undertaken by a number of trained community workers.

Despite the small-scale nature of these two experiments, they highlight some of the problems and potentialities of community development. Truly community-based development requires not only greater community participation and action but also structures sensitive to the local people and their needs. Both experiments provide evidence in support of the potentiality of 'bottom-up' rural development, though in each locality there was little evidence of significant local initiatives. In other words, it would appear that even community-led development can only really be effective if conceived within the context of some institutional structure. Certainly some of the lessons learnt by these two projects are being incorporated into the policies and procedures of several of the sectoral and regional agencies responsible for rural development.

The Rural Development Commission and community-based development

The Rural Development Commission provides an example of an integrated rural development agency that has attempted to respond to the pressure for a greater degree of community involvement in the development process. In one sense, the commission has been ahead of this movement through its funding, since the 1920s, of Rural Community Councils (RCCs) in the English counties. RCCs have been concerned with encouraging voluntary agencies and local communities to become more active in the development process. In recent years field-workers have provided advice and guidance to groups concerned with development issues (Moseley, 1985), and numerous communities have been stimulated to organize a detailed appraisal of their own resources and future needs (Sulaiman, 1988). Even though these 'village appraisals' vary considerably in format and content, they provide an insight into the nature and vibrancy of a local community. Of course, whether a development agency or a planning authority takes any notice of such locally designated needs is another matter.

A more formal initiative to incorporate 'local' involvement into the structure and workings of rural development policy and action has been taken in the commission's 1984 Rural Development Programmes (RDPs). These are broad-based strategies to revitalize the local economy and society of designated Rural Development Areas (RDAs) in the English counties (Green, 1986). Apart from support for the RCCs, for the last two decades the commission has placed most emphasis on the promotion of employment provision in rural areas (see Chapter 7). One element of this work involves the Council for Small Industries In Rural Areas (COSIRA, re-organized in 1988 as the Business Service within the new Rural Development Commission), which has traditionally concentrated its efforts on providing technical and business advice, skill training and limited financial loans to small firms in rural areas. Second, the RDPs and RDAs have a role in helping the commission to target its limited finances on the most needy areas within England. This strategy started in the late

1960s with a small number of 'Trigger Areas' that were later broadened into Special Investment Areas and extended further to 27 RDAs by 1984 (Figure 11.1).

It should also be noted that the commission has chosen to work closely with the local authorities on the RDPs, so that as local authorities have redefined

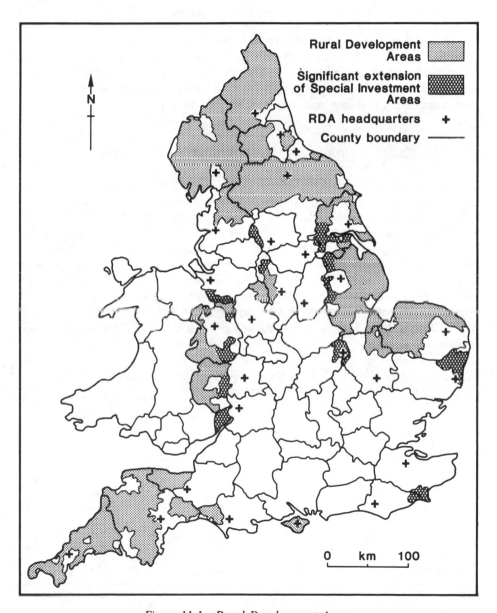

Figure 11.1   Rural Development Areas

their views on the meaning of rural development, the commission has broadened
its approach from the economic to the wider social and cultural development of
remoter rural communities. The new commitment is evidenced, on the one
hand, by the rising share of expenditure on programmes unconnected with
business premises (Table 11.1) and, on the other hand, by schemes of funding
that extend the influence of the commission into such areas as rural transport,
housing and retailing.

Clearly the RDP initiative has considerable implications for the future
direction of rural development strategy. For this reason the remainder of this
chapter reviews and evaluates the operation of these programmes in so far as
they appear to offer scope for a community-based strategy of development. A
key issue to be examined is the extent to which rural communities play a part in
the formulation and implementation of the resulting RDPs.

Place of communities in the administrative structure

Each RDA sets out the development of its RDP through a written document,
which is submitted annually to the commission for review. The documents
identify the particular problems faced in each local area, the strategies being
adopted and the bids made to the commission for financial support for proposed
actions. Turning first to the administrative structure of the RDPs, a review of the
documentation, supplemented by interviews with project and relevant planning
officers in the RDAs, reveals a hierarchical control function between the
commission and rural communities (Figure 11.2). The commission retains
financial control of the development process: all RDPs, and their associated
expenditures, have to be approved by London-based officers of the Commission;

Table 11.1   Development Commission expenditure, 1984/5–1988/9 %

Subject area	1984–5	1985–6	1986–7	1987–8	1988–9[1]
Factory premises	54.0	52.0	50.0	48.0	37.0
Redundant buildings	5.0	6.0	4.5	6.0	8.0
Partnership projects	2.6	1.6	2.3	2.7	2.5
COSIRA operations	24.0	21.0	22.0	22.0	33.0[2]
COSIRA lending	—	1.6	1.2	—	
Administration	3.8	3.3	4.1	5.0	
Social/cultural/welfare	8.0	8.0	9.0	10.0	11.0
Housing	1.7	0.7	0.6	1.0	0.7
Other developments	0.3	1.3	2.3	2.4	4.7
Taxation	—	4.0	2.9	2.0	2.4
Total gross expenditure (£m)	22.4	28.5	25.9	34.6	36.7

Notes 1. Revised basis for allocating expenditure.
2. Includes publicity, research and loans to small firms.

(Source: (Rural) Development Commission annual reports.)

indeed, there is some informal evidence that the DoE and the Treasury influence decisions over the approval or non-approval of certain proposals. On the one hand, this approval system allows the commission to steer scarce funds into RDAs exhibiting particularly severe social and economic problems, especially as measured by levels of unemployment. In practice this has favoured RDAs in the north of England, especially Cleveland (an 11.5 per cent unemployment rate in January 1989 against an average for England of 6.9 per cent, and 6.2 per cent in all RDAs) and Durham (8.8 per cent). It should be noted that South Yorkshire (14.6 per cent unemployed), Cornwall (9.2 per cent) and the Isle of Wight (9.7 per cent) are excluded from this 'enhanced RDA' status. On the other hand, the allocation of relatively small amounts of money are subjected to the same bureaucratic processes of scrutiny, delay and approval as large projects. Except for limited economic and social funds devolved to local managers, the commission has resisted passing to the managers of each RDA responsibility for approving expenditures, with resulting delays and lack of flexibility to respond to local needs and initiatives. The justification for these controls, however, is that public funds are being allocated and some procedure of accountability is required.

Each RDA is managed by a Joint Steering Group (or Strategy Committee) comprised by county representatives from four 'core' organizations: the elected county and district councils covered by the RDA, COSIRA (Business Service), English Estates and the county RCC. The RCC provides representation for voluntary groups on the Joint Steering Group but is an indirect representation as far as individual rural communities are concerned. Some RDAs have broadened the base of their representation by extending membership to other organiza-tions. For example, 26 per cent of RDAs include representation from National Park Boards, 22 per cent involve the Regional Tourist Boards, while the Agricultural Development and Advisory Service of the Ministry of Agriculture, Fisheries and Food, as well as voluntary organizations, are represented on 19 per cent of Joint Steering Groups. But direct community representation, for example, through parish councils, is found in only 15 per cent of RDAs. In broad terms, the Joint Steering Groups exert local political control over the RDPs through the influence of elected county and district councillors, while officers of the commission may also attend meetings to provide guidance and advice.

While Joint Steering Groups make broad strategic decisions, day-to-day management and development issues are left to Officers' Coordinating Groups (Figure 11.2). Given the close working relationship with the planning structure, each group is serviced mainly by personnel drawn from the local authority involved in planning and delivering services and infrastructure in the RDA. Usually a member of the planning department is designated the group's executive officer, although approximately a quarter of the RDAs have appointed a specialist project officer. The latter tend to have their offices located within the RDA, whereas the former, with other duties to perform, retain their offices in the county town, often at some distance from the RDA they are administering (Figure 11.1). Officers' Co-ordinating Groups determine the

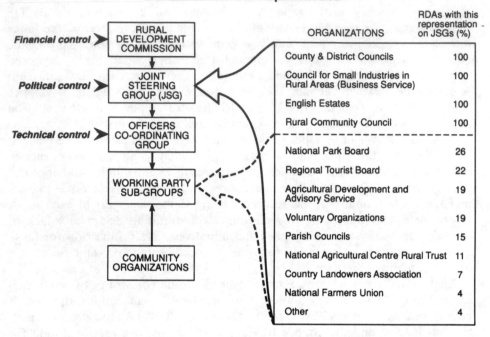

Administrative Structure
of each Rural Development Area

	ORGANIZATIONS	RDAs with this representation on JSGs (%)
	County & District Councils	100
	Council for Small Industries in Rural Areas (Business Service)	100
	English Estates	100
	Rural Community Council	100
	National Park Board	26
	Regional Tourist Board	22
	Agricultural Development and Advisory Service	19
	Voluntary Organizations	19
	Parish Councils	15
	National Agricultural Centre Rural Trust	11
	Country Landowners Association	7
	National Farmers Union	4
	Other	4

Figure 11.2   Administrative structure of each Rural Development Area

Table 11.2   Strategic objectives

Subject area	RDAs (max. 27)	%
Employment	27	100
Transport – accessibility	27	100
Information and advice	27	100
Community care	27	100
Community facilities	27	100
Housing	27	100
Job training	26	96
Land and factory premises (workspace)	25	93
Shopping	24	89
Tourism	24	89
Support for businesses	22	81
Recreation	22	81
Education	21	78
Arts	16	59
Environmental improvement	14	52
Health	11	41
Transport – roads	18	27
Marketing	5	19

(*Source:* Rural Development Programmes of the RDAs 1986/7–1988/9.)

technical feasibility of proposals under RDPs and ensure their conformity and implementation within county Structure and Local Plans.

Proposals on actions, as well as detailed developmental work, emanate from Working Party Sub-Groups (Figure 11.2). Interest groups can have their greatest representation and influence at this level since the working parties are able to receive proposals from individual community organizations. Consequently, while RDAs perpetuate a 'top-down' structure for rural development, there is scope for 'bottom-up' community-led development, albeit heavily mediated by layers of technical, political and financial control. In addition, the designation of an RDA within a local authority serves both to focus the attention of local policy makers on the problems of the rural area and to provide a management structure to foster consultation and co-ordination between the various sectoral agencies operating in that area. However, it should be noted that the RDP institutional structure is non-statutory, a circumstance that is used to deny the local Joint Steering Groups power to spend public money.

Attention to community development in RDPs

The objectives

One measure of the extent to which the community interest is recognized within RDAs can be taken by examining the stated objectives of the 27 RDPs. A review of the programmes shows a common core of subject areas – employment, transport, housing, information/advice, community care and community facilities (Table 11.2). To this list can be added a number of less popular, but still important, objectives such as for job training, land/workspace, tourism and shopping facilities. In each case there is a commitment to developing the provision or performance of facilities within the subject area. While economic considerations predominate, the community interest is explicitly represented in two objectives – community care and community facilities – and indirectly in three others – advice/information, accessibility and shopping. At the other extreme, a number of objectives are weakly developed within the RDPs; environmental improvement, for example, is a stated objective in only 52 per cent of cases, while the arts (59 per cent), health (41 per cent), roads (27 per cent) and marketing (19 per cent) are also poorly represented. Support for businesses, education and recreation lie between these extremes.

Given the common criteria by which RDAs were established, it is not surprising that the formal objectives of the RDPs exhibit a high degree of similarity. This feature is also encouraged by the periodic conferences organized by the Rural Development Commission for the executive officers of the RDAs, at which the development of RDPs is discussed. Nevertheless, the interests of a number of groups within the rural community are under-represented in the objectives of the RDPs; these groups include the elderly, women, the handicapped and the unemployed. While it may be claimed that the needs of these disadvantaged groups are met within the context of the broad strategic objectives, the absence of explicit objectives can allow their needs to be overlooked.

The projects

At all levels of development planning, a dissonance commonly exists between the good intentions expressed in programme objectives and tangible development projects. Consequently, a second measure of the extent to which the community interest is recognized can be taken by examining the bids actually made to the commission for financial support under the RDPs. Table 11.3 shows the pattern of bidding for the three years 1986/7–1988/9. The bids are categorized under headings that allow some comparison with the strategic objectives listed in Table 11.2. The continuing predominance of the 'economic' interest is evident, with approximately 40 per cent of bids in each year tied to the provision of 'factory premises'. Even so, 'community development' is the next most important category, accounting for around 25 per cent of all bids over the 1986–8 period. However, all of the remaining objectives are poorly represented in the bids actually made under the RDPs, even though the rising proportion of bids under 'tourism' and 'environmental' is significant. One problem here lies in the prior existence of other agencies, often with national responsibilities for development in a particular subject area (sometimes referred to as the 'domain'). The Manpower Services Commission (now the Training Agency) in relation to job training, the Department of Transport in the area of road provision and the DoE for housing policy, exemplify this problem of inter-agency competition. Such competition commonly excludes meaningful activity by individual RDAs in certain policy areas, even though the failure of these other agencies prompts the need for local action. From this perspective, the Rural Development Commission, and its RDA programme, is restricted to operating in the interstices between the domains of the large, centralized departments of State and their dependent agencies: thereby RDPs are largely excluded from many areas of policy that are fundamental to rural development, notably housing, job training and road provision.

Not all bids succeed, however, and the constraints on Rural Development Commission policy can be seen in the proportions of bids actually approved for funding (Table 11.3). Proposals for factory premises, project officers, information/advice centres, village halls and community development show consistently high approval rates. By comparison, variable, and in some cases falling, levels of funding are evident in sectors where inter-agency competition exists, notably in job training, tourism, housing and infrastructure.

A third component in translating objectives into tangible projects lies in the proportion of the total cost the commission is prepared to fund. Given the very limited budget supplied by central government – only £36.7 million in 1988–9 – much emphasis is placed on the 'pump priming' role of the commission: most of the finance has to be obtained from alternative sources, including local authorities, other agencies, the European Community, charities, trusts, voluntary bodies and the private sector. A 'leverage ratio' can be calculated as the

Table 11.3   Rural development projects, 1986/7–1988/9

Subject area	Number of bids (%)			% bids approved*			DC grant as % total cost		
	86–7	87–8	88–9	86–7	87–8	88–9	86–7	87–8	88–9
Factory premises	31.0	41.0	40.0	88	96	79	—	100	96
Tourism	11.0	11.0	13.0	56	77	25	18	14	12
Job training	1.8	0.5	1.6	20	83	32	50	13	7
Small businesses	2.7	2.8	1.4	48	81	18	33	49	12
Housing	5.1	5.1	2.9	28	97	3	22	8	0
Social and rural services:									
Community development	28.0	24.0	22.0	76	93	94	19	15	20
Information and advice	6.1	1.6	2.5	71	100	97	77	59	49
Village halls	6.8	10.0	7.3	86	98	93	22	17	17
Other projects:									
Project officers	0.6	0.5	1.6	100	100	100	50	44	38
Promotion	0.8	—	—	57	—	—	66	—	—
Environmental	1.7	1.1	3.4	79	77	15	22	20	23
Infrastructure	1.1	0.8	—	0	67	—	9	11	—
Other	3.0	1.7	6.5	64	70	51	17	43	41
Total (number)	(841)	(1146)	(1189)	72	92	69	21	14	48

Notes *The proportion of projects given a formal offer of a grant is lower.

(*Source:* (Rural) Development Commission annual reports.)

ratio of commission funds to financial support from elsewhere. In 1985–6, for example, the overall leverage ratio was 1:3.3 – just over £3 for every £1 of commission money. This ratio rose in successive years to 1:4 and 1:66, but fell back in 1988–9 to 1:4.3. An examination of the sources of external funding on a project-by-project basis reveals the predominance of local authorities, the relatively small private-sector involvement overall, and with particular agencies collaborating in each subject area. The private sector and the English Tourist Board, for example, are important in the funding of tourism projects, whereas the Housing Corporation is usually involved in the relatively few housing projects under development. The leverage exerted by the commission within these partnerships varies considerably (Table 11.3). Factory premises, for example, are almost totally commission-funded, whereas bids under community development have to attract 80 per cent of their finances from other sources. Table 11.1 has already shown the small, if growing, percentage of commission funds allocated to social/cultural/welfare developments. The contribution of commission funds in other subject areas ranged in 1988–9 from 49 per cent in developments concerned with the provision of information/advice, through 38 per cent for project officers and 23 per cent on environmental projects, to 12 per cent on tourism developments. The requirement that non-commission funding be generated clearly acts selectively on the bids made, on their approval rate and consequently on the projects eventually implemented in the RDAs. The aspirations or needs of a rural community, therefore, are mediated not only by the hierarchical management structure described in Figure 11.2 but also by the willingness of a number of agencies to collaborate in the design and funding of a particular project.

The content

An examination of the 27 RDPs, over the first three years of their existence, reveals a wide variation both in the types of projects and the vigour of their development. Because there is little commonality, three RDAs have been selected to illustrate the content of RDPs. The projects can be gathered under three headings: economy, services and community (Table 11.4). In the Suffolk RDA, projects concerned with services have been particularly well developed, whereas the Hereford RDA lays more emphasis on economy and community. In East Cleveland, the services sector is also less well-developed, as are projects under the community heading. Projects range widely from the provision of mobile services, such as cinemas, libraries, advice centres and playbuses, through the development of infrastructure on housing, village halls and bypass relief roads, to very specific community projects such as a welfare-rights officer, a musicians' collective, canal restoration and an allotment co-operative. While the lists in Table 11.4 emphasize the variety of projects under development, there are some common features. For example, the construction of workshops, the conversion of premises for workspace and the provision of council houses, libraries and community centres are common to all RDPs. Indeed, a greater degree of convergence can be expected in the longer term, once the Joint

Table 11.4　The content of Rural Development Programmes

Project area	Rural Development Area		
	Suffolk	Hereford	East Cleveland
Economy	Workshops Conversions Land purchase Site servicing Bypass relief road Transport broker	Workshops Advance factories Conversions Land purchase Marketing initiative Rate concessions Small business training 　programme Bypass relief road Car parks Transport broker	Workshops Conversions Economic- 　development officer Crafts co-operative Training and 　enterprise centre Skills promotion and 　management 　courses Tourist development 　feasibility study Heritage Centre
Services	Shared-equity scheme Council housing Sheltered bungalows Housing renovation Mobile library Mobile cinema Educational project Theatre touring group Information officer Citizens Advice 　Bureau Shopkeepers 　association Community bus Museum	Share-equity scheme Council housing Sheltered housing Library Mobile cinema Community bus Museum	Council housing Library for teenagers Library for elderly Citizens Advice 　Bureau Village arts
Community	Village halls Play areas Mobile community 　centre Home-aid teams Community 　development 　courses Multi-games hall Village appraisals Volunteer centre	Village halls Community centre Information/advice 　centres Community 　development courses Playbus scheme Community association Housebound-reader 　service Parish-council training Countryside-awareness 　course Canal restoration Picnic sites	Community centre Sports recreation 　centre Welfare-rights officer Book craft and play 　stimulus Community worker Allotment co- 　operative Business's collective Community centre and 　support team

(*Source:*　Rural Development Programmes of each RDA, 1986/7–1988/9.)

Steering Groups become aware of developments in other RDAs and are able to adopt initiatives being implemented elsewhere (Development Commission, 1984a). Even so, each RDP reflects local problems, opportunities and perceived needs, as well as the willingness of agencies to collaborate within the RDA.

The role of rural communities within RDAs

An insight into the role of rural communities within RDAs has been gained through unstructured interviews with a sample of the executive and project officers. The interviews were not designed to produce quantitative data but rather to be exploratory in nature and so provide the basis for further community-based research.

An important distinction can be drawn between development projects that benefit the whole RDA and those specific to particular communities – the latter defined in terms of one or a group of parishes, villages or small towns. 'Broad benefit' projects tend to be initiated by the development agencies, such as COSIRA (Business Service), or a department within the local authority, for example, as regards job training centres, the development of a museum or the provision of a transport broker. Here the 'top-down' model of policy formation is descriptive of the process of rural development. 'Specific benefit' projects, on the other hand, tend to be initiated by particular rural communities and come close to the 'bottom-up' approach to development. In these cases, particular local needs are advanced by local interest groups, private individuals or elected representatives. The resulting developments include community buses, heritage centres, tourist developments, sports and recreation centres and village appraisals.

When the locational distribution of the 'specific benefit' projects is examined, the dispersed and uneven impact of RDPs is revealed. Figure 11.3 shows the allocation of some of the proposed projects for two RDAs – Lincolnshire and North Yorkshire – over the first three years of their RDPs. Developments are localized to the benefit of certain communities but with many villages and parishes receiving no project developments. No doubt these spatial biases can be corrected in the longer term, but this assumes the continuation of the present RDA strategy of the Rural Development Commission for five years or more. To date, the community-led initiatives appear to be benefiting the larger villages and small towns, with no assurance that the pattern of developments yet reflects the 'needs' of the communities within an RDA.

Conclusion

Community-led rural development is still in its infancy in the UK and at present has to operate in the interstices between the spheres of influence of pre-existing sectoral and regional development agencies. Indeed, rather than emerging as an alternative model for development, the 'bottom-up' approach seems most likely to be absorbed by the established institutional structures. For example, the RDA strategy of the Rural Development Commission still confirms the dominance of the 'top-down' approach to rural development, albeit with some

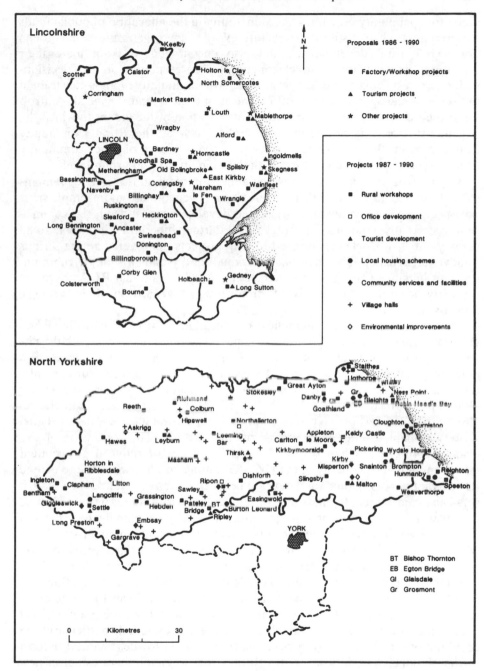

Figure 11.3   Proposed projects in Lincolnshire RDA and North Yorkshire RDA

scope for community initiatives. Accountability in the allocation of public funds is normally used to justify this structure. While new organizations have been drawn into rural development, and also encouraged to co-ordinate their operations in local areas, direct community representation in the decision-making process remains weakly developed. Any community input is constrained by several managerial layers of technical, political and financial control. Nevertheless, there is scope for further investigation of the extent to which the community 'voice' is represented by bodies such as the Rural Community Councils on the Joint Steering Groups, and the influence of other community agents, such as parish councillors, at the working-party sub-group level.

The analysis has also shown how the allocation of commission funds remains strongly biased towards workspace provision, even though more emphasis is now being given to community-orientated developments. Indeed, the commission has recently announced a 'Rural Social Partnership Fund', applicable to all rural counties in England, designed to help projects that overcome the serious disadvantages experienced by some individuals and groups within the community. This initiative goes some way to meeting the objection that RDAs have not explicitly identified the needs of disadvantaged groups in their strategic objectives.

This latest initiative exemplifies the flexibility with which the Rural Development Commission attempts to deploy its limited resources, including its 'leverage' role in extracting additional funds from other agencies. There are a number of initiatives outside the RDA programme of this nature, for example, the Rural Transport Development Fund. Nevertheless, Hoggart and Buller (1987, p. 176) conclude that all of these measures represent 'conscience-absolving palliatives for the problems of peripheral rural localities rather than a serious attempt to alter their socio-economic setting'. The conclusion of this chapter, however, is less severe. In a political era of reduced government involvement in the economy of the UK, the commission has been able to extend its programmes of regional rural development, and offer a stimulus, even though small and constrained, to innovative development projects in rural communities.

Nevertheless, for 'bottom-up' or community-led development to achieve a wider role within RDAs, a significantly greater proportion of commission funds will need to be directed away from the emphasis on workspace provision and towards wider community developments, including project and field officers working at the community level. There is now a broad-based recognition that 'employment deprivation' is a necessary but not sufficient focus for contemporary rural development. In addition, the commission will require greater direct access into policy areas from which it is at present excluded, particularly the provision of rural housing. At present the commission provides financial support to the National Agricultural Centre Rural Trust to enable it to establish new rural housing associations; but the influence of the commission on rural housing provision is indirect. Unfortunately, the prevailing political climate does not favour either type of policy revision. Consequently, community-led development in the RDAs seems likely to retain its small-scale, uneven impact on the problem rural areas of England.

From the evidence reviewed in this chapter it would appear that the dynamics of rural communities in promoting their own interests has considerable potential for further initiatives in community development. Yet, despite this, potential community-based development still remains small scale in rural Britain. It is clear that if this approach to development is to be adopted more widely then there is need for those involved to take account of the lessons learnt by community-based projects including those in the Leominster Marches and Peak Park and elsewhere. Within any successful scheme of community development the role of 'internal' community leadership in stimulating positive attitudes towards development has to be recognized – as does the part played by external project officers and field-workers. Such individuals can stimulate community groups towards developing entrepreneurial attitudes, identifying local needs and formulating development proposals. Such development is inevitably labour intensive and a high financial cost strategy (Parker, 1984). There is also a need to shift from working 'through' the community to working 'with' a community, thus emphasizing the emergence of policies from the community as well as the involvement of all its members. To achieve this end it is vital that structures are created that are sensitive to local people so that greater involvement in the development process is achieved. Without such changes community development will remain a mirage.

Acknowledgement

The authors wish to thank Professor Michael Chisholm and Mr James Derounian for their comments on an earlier draft of this chapter.

12

SOURCES FOR THE STUDY OF SOCIAL CHANGE IN THE COUNTRYSIDE: THE NEED FOR INTEGRATION

Nigel Walford and Ann Hockey

A major theme of this book has been its emphasis on social change. Chapter 1 stressed the fundamental nature of recent changes in the countryside, showing that they have stemmed in part from the arrival of new types of people and activities in rural areas but also that they arise from wider structural changes in society. Subsequent chapters have documented the reversal of population and employment trends in many parts of rural Britain and have investigated the impact of changes in housing, labour demand and service provision on the different types of people living in the countryside. Some of these developments represent the mirror image of what is happening in the more urbanized parts of Britain, most notably the redistribution of people and jobs, while others (such as the contraction of rented housing and concentration of services) appear to follow national trends quite closely. The general effect, however, seems to be a further blurring of the distinction between town and countryside, seen perhaps most vividly in the convergence of industrial structures (Chapter 6). Nevertheless, life in the countryside retains a high degree of distinctiveness because of the very different geographical context resulting from generally low population densities and thin spread of jobs, shops and other facilities. This is, of course, particularly evident in such an extremely peripheral and rugged region as the Scottish Highlands and Islands (Chapter 10), but has also been seen to be true in much less remote parts, such as Suffolk (Chapter 7), the Cotswolds (Chapter 8) and North Yorkshire and Lincolnshire (Chapter 11).

This contrast between the continuity of basic geographical context of the countryside and the changes taking place in the demographic, economic and social characteristics of rural areas provides the central theme of this final chapter. On the one hand, we are forced to recognize that, because of the changes, the study of people in the countryside has increased considerably. At one time many demographic studies of social change in the countryside were frequently subsumed under the investigation of population change, but since the 1960s interest in social change – led notably by Pahl's *Urbs in Rure* (1965) – has developed beyond consideration of people simply as inhabitants of rural space towards viewing them as social actors in the economy and society of the countryside. The result of this increased concern for social change is that new questions are being asked about people in the countryside. It is no longer sufficient to measure the size of the rural population and look at its age and sex structure. There is a need to relate such demographic parameters to the social characteristics of the inhabitants of the rural areas and to the social and economic networks that exist there.

On the other hand, in trying to do this we are faced with a set of methodological and technical problems that have changed very little. These relate essentially to the low density of the population and to the large number of small settlements found in the countryside. Purely for statistical reasons let alone because of the real diversity that exists within and between rural areas in Britain, it is harder to generalize about people in the countryside than about the characteristics of large towns and cities. Indeed, it is extremely difficult to obtain access to statistical data for individual rural settlements from standard national sources. This is even true of the basic counts in the Population Census, where the parish may be the smallest statistical area for which data are available, but the task is virtually impossible when it comes to more sophisticated measures of social composition and to monitoring trends between censuses. It is no coincidence that rural studies tend to rely on case study work to a greater extent than research in urban geography. Indeed, at the end of this chapter the reader may well consider that the emphasis on 'case studies' in this book is doing no more than making a virtue out of necessity!

In this context of change and continuity, this chapter considers whether the existing statistical sources are capable of fulfilling the new requirements placed on them – essentially whether they have developed in parallel with the changes that have taken place over the last twenty-five years in the types of questions being asked about people in the countryside. Its primary aim is to review the data sources available to students of the subject. Those sources that have typically been used for rural population research are reconsidered, and new sources are examined that might fulfil the emerging demand for a different kind of information base in the light of increased emphasis on social change. In particular, the complementary roles of national and local sources are investigated and it is suggested that although care must be taken, their integration may offer the opportunity for more penetrating analyses. A subsidiary aim of this chapter is to provide more information for those who are interested in following up some of the trends observed in previous chapters or in undertaking similar studies in their localities. For these people, the chapter not only provides more detail on particular data sources than provided in previous chapters but also indicates the potential and the weaknesses of using these sources for the study of rural change.

National and local data sources

A key question is the extent to which the diverse sources of data are capable of being linked together in order to complement each other. One important point related to the opportunities of data linkage is the extent to which different data sources are based on incompatible spatial units or alternative definitions of rural areas. The question of data linkage has become particularly important in the light of the 'information revolution' spreading through many walks of life. The collection of information by local organizations, especially local-government departments, has the potential to allow centrally collected, national sources to be integrated with locally derived material.

The distinction between nationally and locally derived information provides a convenient framework for examining different sources in respect of their suitability for investigating population and social change in the countryside. Table 12.1 summarizes the different national and local sources considered in this chapter. National sources are taken as those that are either collected by an organization (e.g. a government department) that has a national remit, or are collated on a nationally consistent basis, although they might have been collected locally. Local sources in contrast are collected for sub-national areas, typically local-government areas such as counties or districts, generally by organizations with a remit relating to the area.

Table 12.1   National and local sources for rural social change

	Typical small spatial unit	Estimated % of counties*
National		
1. Population Census	Enumeration district	
2. Vital Statistics	Ward	
3. NHS patient re-registration	Family practitioner Committee area	
4. Mid-year Population Estimates	Local-government district	
5. Social and economic surveys	Standard region	
Local		
1. School roll	Parish or ward	30
2. Electoral roll enhancement	Ward	15
3. Rural facilities	Parish	65
4. Household surveys	Ward	38
5. Population estimates and projections	Parish or ward	47

Note * Based on a survey of county-council planning departments in England and Wales in 1986–7.

National sources

The Population Census, which has been carried out every ten years since 1801 (with the exception of 1941), is generally regarded as the most useful source of information about the population. Its value has been enhanced in recent decades through a number of important developments. First, an abundance of detailed local-level information has been published as a standard set of cross-tabulations known as the Small Area Statistics (SAS). These SAS are a consistent source across the whole country and are produced for various spatial units including enumeration districts, civil parishes and wards.

Second, although the content of the Population Census is considered to be rather restricted by some researchers, these SAS are extremely detailed cross-tabulations of the full range of Census questions. Thus they include information that helps in answering questions about social as well as simply population change. One criticism that has been levelled at the SAS is their inflexibility, although the Census authorities counter this by asserting that special tabulations can be obtained if the predefined SAS cross-tabulations do not meet users' specific requirements.

Third, the availability of these SAS and other related ancillary Census material to academic researchers through the Economic and Social Research Council Data Archive at the University of Essex has encouraged their widespread use. This service has enabled the computerized transfer of these statistics to be carried out relatively easily, without the need for researchers to purchase or computerize the statistics themselves. Fourth, the Census is seen as a regular source of compatible information, albeit one that has a ten year time interval. Changes to the content and the geography of the Census take place reflecting new information requirements, but these can sometimes be overcome or minimized by the judicious selection of areas and variables. Other writers, for example Rhind (1983) and the Census authorities (OPCS 1981a), have described the content and definitions used in the Population Census.

A fifth development that occurred following the 1981 Census was the general provision of small area-level statistics on migration and population characteristics for areas as places of work. Although these additional sources, especially the former, suffered from restrictions on data availability due to confidentiality, they did enable some new research questions to be addressed. For example, the Special Migration Statistics provided some details of the social characteristics of migrants to and from small areas, such as wards or postcode sectors. Unfortunately because a substantial number of wards in rural localities had comparatively low numbers of in- or outmigrants, the matrix of migrations between wards was rather sparse, leading to data suppression.

Finally, the Census statistics for small areas can be linked into, or have been used for producing, definitions of rural areas (Cloke and Edwards, 1986; Craig, 1988). The Census authorities and other government departments have been aware of the need to examine changes in rural areas, and for that matter urban ones. As a result 'official' definitions of rural areas have been produced and reports documenting the extent of demographic, economic and social change of these areas have been published. Furthermore, these definitions of rural Britain are available for researchers wishing to use them in their own work.

At the present time sights are being set on the output from the next Population Census, which is due to take place on 21 April 1991. As already indicated, one of the main strengths of the Population Census is that it provides reasonably compatible statistical information at regular intervals. Nevertheless, opportunities are taken for improving or enhancing the output. In the preparation for the 1991 Population Census, new topics have been considered for inclusion, for instance, long-term illness, dependency and 'concealed families', and a common statistical base will be created for use in the various forms of published material (Clark and Thomas, 1990). The first of these new topics, long-term illness, is likely to be of particular interest to those concerned with the elderly population in the countryside. Furthermore, developments in the area of geographical information systems and digital cartography are likely to mean that the links between Census and other sources of data will be improved.

The annual Vital Statistics on births and deaths by age and sex from the Office of Population Censuses and Surveys are another important source for rural population research. These statistics are generated through the system of registration of births and deaths and are issued for local-government wards. Although the actual content of this source is rather limited, recording nothing about the social background of the deceased or the parents of the babies, it is useful in enabling levels of population to be measured. The Vital Statistics allow for the natural increase component of aggregate population change to be calculated. This source therefore makes a significant contribution to research that attempts to discover where rural repopulation is taking place. Since these statistics are produced for wards it is possible to link them into other information sources. This is essential as the content of this source is exclusively demographic. It is only by associating these data with other information that one can go any way towards looking at, for example, indicators of mortality in relation to socioeconomic conditions.

The National Health Service central patient re-registration system is a continuous system that may be used for monitoring migration. Each time that people change their GP this re-registration is recorded on a central computer system. When re-registrations involve a move between Family Practitioner Committee (FPC) areas then the statistics produced by the system can be used to estimate migration. There are certain limitations on the usefulness of this source. First, it has been shown that there are inconsistencies between different groups in the population in the time taken to re-register with a GP. This can lead to bias in the estimates of migration. A second problem is that the FPC areas are fairly large for instance, being the metropolitan districts and non-metropolitan counties in England and Wales. This means that migration patterns revealed by the statistics are at a fairly crude level and not necessarily suitable for examining migration between urban and rural areas. The coarseness of the spatial scale used in the data gives rise to a third difficulty. Short moves across an FPC boundary count as a migration, whereas potentially longer moves within the same FPC do not. For a long time there has been debate about when moves by individuals or households should be counted as migration and when as local circulation (Jones, 1981). When the data source is as crude as in the NHS registration data, the problem of definition is compounded.

The national social and economic surveys conducted by government, principally the General Household Survey (GHS), the Family Expenditure Survey (FES) and the Labour Force Survey (LFS), contain the types of information needed in order to go further than simple population counting. Their major drawback is that their only geographical reference is the standard economic planning region. The samples used in these surveys are representative at the national and regional levels. However, they would become unreliable and prone to breaches of confidentiality if information was to be released for geographical units below the standard region.

The GHS, for example, permits the construction of marriage-to-birth intervals and enables these population events to be associated with the migration, personal circumstances, employment and economic conditions of households in the different regions, but these interesting research questions cannot be probed at a lower geographical scale. The GHS is an annual survey and for a period in the 1970s, it included a variable relating to the household's location in terms of urban or rural area, based on the type of local authority. Although this only gave a very broad and geographically unspecific impression of social change in rural areas, it was potentially useful to researchers as a means of distinguishing in a general sense between the social conditions of 'rural' and 'urban' households.

The FES allows researchers to explore in great detail the income and expenditure patterns of individual households in association with information about employment, housing and social status. Apart from the inclusion of a classification of the sampled households according to the population density of their residential district, the survey suffers from the same lack of geographical detail as does the GHS. These national surveys can unfortunately really only provide a context in which to place the more spatially specific information contained in such sources as the Census and the Vital Statistics for the whole country and the local data discussed in the next section.

Local sources

Local information sources are as diverse as the organizations responsible for collecting the information. An indication of the range of computerized information relating to the countryside is provided in the Rural Areas Database Catalogue (Walford, Shearman and Lane, 1989). Rather than attempting the near impossible task of describing all the different surveys and administrative sources conducted or collected locally, this section identifies the major types of local source available. This will therefore serve as a guide to what might be available for particular areas of the country.

Broadly speaking there are two main types. The first group is composed of sources often derived from research studies exploring a particular issue, for instance, the social characteristics of inmigrants to a given area and their reasons for migration. From such a source it may be possible to generalize in order to make limited claims about the broader rural population. The great variety and specificity of sources contained in the first group means that it is inappropriate to do more than mention one example. This is the Rural Deprivation Study carried out for the DoE in 1981 (McLaughlin, 1986). Although this study's geographical coverage was limited to five clusters of parishes in different parts of England, attempts were made to relate this local source to compatible national statistics from the Population Census. In many ways this survey constitutes the definitive statistical source on social conditions in rural areas and it would be appropriate for a repeat survey to be undertaken in order to generate more up-to-date information and to enable social change to be investigated.

The second group of local sources comprises those that duplicate or update information gathered at the national scale together with additional variables. It also includes those sources not necessarily collected by the majority of local organizations but that are sufficiently common not to be regarded as 'one-off' surveys. Sometimes these sources are the more detailed versions from which national statistics are compiled. Examples of this group of sources are discussed below, and include sample household surveys and local population estimates and projections. The continuing need for accurate and up-to-date information on the rural population is a contributory factor in prompting many local authorities (districts and counties), sometimes in conjunction with rural community councils, to collect social and demographic information relating to their areas.

There are five main, commonly occurring types of local survey or administrative statistical source, relevant to examining social change, that are collected by local authorities: rural facilities surveys; general purpose household surveys; school-roll records; electoral roll enhancements; and population estimates and projections. Other types can be identified, but this list represents a set of sources that are conducted fairly widely. Although there are inevitably differences in detail between, for example, a rural facility survey in one county or district compared to another, there are sufficient common elements present to enable a broad picture of change to be constructed.

The rural facilities surveys now carried out in many counties and districts have originated in response to a need for information about the acknowledged problem of declining services and facilities in rural areas. In many instances these surveys have been undertaken as part of a broader review of rural areas and for the purpose of assessing supply levels of various facilities provided by the public and private sectors. The surveys usually contain details of the population and when they have been carried out on a regular basis can form an important basis for researching into various aspects of social change.

A few county councils have been conducting such surveys for many years, such as Norfolk, which has had a rural facilities survey since 1950, but in most authorities the commencement of their series of such surveys dates from the 1970s or 1980s and appears to be connected with various factors. The introduction of structure planning responsibilities for counties as part of local-government re-organization in 1974 prompted the need for authorities to have more information about the rural residents. There was a realization of the consequences of protracted depopulation – in particular, a concern about the loss of services for the remaining population as a whole and for certain sectors of the population especially. Also the general spread of the 'information technology society' meant that the collection of statistics became part of the 'administrative machinery'. The decline in rural services has been the subject of research by social geographers for some time and is seen as part of the general demise of 'village life' (Standing Conference on Rural Community Councils, 1978; Dawson and Kirby, 1979; Mackay and Laing, 1982).

In addition to surveys about the facilities available to those people living in the countryside, local authorities also collect information relating to the social characteristics of the inhabitants. Although these are not normally as detailed as their national counterparts (the GHS and FES), they do provide information for smaller areas, thus enabling social change in the countryside to be identified. General-purpose social surveys are not as numerous as rural facilities surveys, although the need for the kind of statistics they can provide is becoming more apparent. Wrekin District Council, for instance, has been publishing an annual digest of statistics for the last four years (e.g. Wrekin District Council, 1989) based on the synthesis of secondary statistics with internally conducted surveys. Where local authorities have not themselves undertaken the collection of primary data on rural social change, they may have resorted to integrating material from various other sources in order to construct socioeconomic profiles of their rural areas.

The county education authorities are required, as part of their statutory obligations, to complete and submit statistics on the size of the school rolls within their areas. These statistics are used to compile national- and regional-level education statistics. However, the more detailed ward, parish or postcode sector figures are used by county councils themselves in looking at patterns of and planning for education provision. This material is usually restricted to counts of the numbers of children, with the sexes and sometimes age-groups separately identified. However it is potentially useful to rural social geographers, since the closure of schools in country areas is a topic that continues to be causing concern. Furthermore, the declining numbers on the school rolls in rural areas are likely, in due course, to become a reduced adult population in the countryside.

The fourth type of local information source, electoral roll enhancement (ERE) surveys began to appear in the 1970s. Their potential was appreciated when the proposed mid-term sample Population Census in 1976 was cancelled. Local authorities had been anticipating the arrival of this up-to-date source of population information and then when it was axed at the last moment looked around for other ways of obtaining basic population counts. Dunn and Swindell

(1972) discussed the usefulness and accuracy of this source, which again merits examination in the light of the community charge or poll tax. Another more recent discussion of ERE surveys (England *et al.*, 1985) has indicated that they continue to be important and provide a means for local authorities to enumerate their populations. The enhancement of the electoral roll to include a limited amount of sociodemographic information, even if this is just the age of individuals, is not without difficulties. The first problem is that, for whatever reason, some people do not complete the electoral registration form. Second, it can only yield information about people under the voting age if this is provided by other members of the same household. Another problem of more recent origin is that the public might be deterred from providing extra information in the belief of a connection between the community charge or poll tax records and the electoral registration system.

The fifth type of local source produced by virtually all local authorities are population forecasts and projections. While central government agencies such as the Office of Population Censuses and Surveys and Registrar General for Scotland publish population estimates and projections, these are available in no more detail than district level in most instances, and indeed at county level for projections for England's Shire counties. However for local planning purposes more detailed figures are necessary. These are sometimes generated by the local authority itself or are increasingly being obtained from the commercial information and computer consultancies that have arisen in response to the general demand for statistical data. The basic population estimates and projections generated by local authorities tend to contain only population counts, possibly broken down by age and sex. However, some of the commercially available material does include area typology information. This 'social area typing', sometimes known as geo-demographics, has usually been produced by means of a multivariate classification of Population Census small-area statistics.

These five main groups of surveys and administrative sources are of course not the full range of information collected by local authorities. Local social-services departments are potentially a very useful source of information on individuals and households receiving social welfare in its various forms, since most will maintain some form of client database. Unfortunately, at least from the researcher's point of view, access to this source will almost certainly be restricted for reasons of confidentiality. The local sources described are those that are more readily available and that suffer less from the problems of confidentiality.

The balance between availability and confidentiality of information is a particularly important issue with regard to investigating social change as distinct from population change. One of the prime objects of research with the former is to probe into those aspects of people's lives that they may not wish to become 'public knowledge'. On the other hand, while a few people object to the Population Census, most accept that there is a need to count the population. The consequence of this problem is that researchers are frequently forced into using sources that are not ideally suited to their particular investigations.

Information integration

One of the purposes of this review of sources for social change in the countryside is to consider the potential for integrating and linking together statistics from different places. The idea of data integration is frequently put forward as one of the major benefits of information technology. However, this claim is often made without due regard to exactly what is possible or desirable. The previous section has given an overview of the principal types of data available both nationally and locally, so this section will consider ways in which some of these could be linked together. Not all of the data sources that have been identified could or perhaps should be integrated in this fashion.

It is important to distinguish between data integration on a geographical or spatial basis and on a substantive basis. The former involves relating data by reference to the spatial or geographical entities for which the statistics are available. One example of this would be linking together aggregate statistics relating to parishes with information on households in specific villages. Substantive integration involves linking together data on complementary or similar topics, for instance, statistics on household income with those about the use of public transport services for the areas in which the households are located. Sometimes both approaches are attempted, but the distinction helps to clarify some of the problems. The problems of data integration fall under three main headings: incompatibility in the geography of the sources; lack of consistency in their content and definitions employed; and variations in the time period of the information.

In theory, and now to some extent in practice, the basic problem of geographical incompatibility between sources is being overcome by the introduction of geographical information systems. This technology is capable of linking together and overlaying data sources collected and computerized for different basic spatial units, for example, grid squares, parishes and wards, as well as a host of more specific types of area designation, for instance, including Rural Development Areas and Less Favoured Areas. However, the ability to perform these kinds of operation on the data is now bringing into focus other related problems. The overlaying of data can produce estimated figures, but if these are to be useful and valid then it is necessary to have a measure of their reliability.

The identification of a lack of consistency between different statistical sources is by no means a new problem. However, it would be easy for these to recede into the background, now that some of the geographical integration problems are being overcome. At a superficial level, it might be assumed that the existence of Population Census statistics for a number of years implies that the data for the separate years are entirely compatible. Unfortunately, this is not the case. Changes in definitions are introduced for some of the variables, for example, comparisons between statistics for the 1971 and 1981 Censuses are hampered by the fact that a different population base was used in the two years (OPCS, 1981a). It does not seem as though the statistics that will emerge from the 1991 Census will be free of this problem either, since it is proposed that an entirely revised occupational classification scheme will be introduced, which will replace the 1981 Census occupational classification and the Classification of Occupations and Directory of Occupation Titles (CODOT) system (Thomas and Elias, 1989). There is also likely to be another new population base introduced for the 1991 Census. These are difficulties with the same source over time: when one looks at different sources for the same time there are many examples of the use of incompatible definitions for the same variable or attribute.

The discussion about the Population Census has indicated some of the problems of integrating the same source over time in order to measure social and population change. Similar difficulties arise in respect of other statistical sources. A number of the county councils that have undertaken rural facilities surveys have now done so on perhaps two, three or even four or more occasions. Although there are usually common elements between the separate surveys, new facilities are included and others are dropped from the questionnaire depending on what are perceived to be the problem services at the time. This is clearly commendable from the point of view of keeping the content relevant, but can be difficult for researchers interested in longer-terms trends.

Two simple examples of data integration have been mentioned, and these represent the two main types currently produced fairly easily. The first is the area profile or compendium composed of information from different sources and presented as a set of statistics for a given area. The Wrekin District Council's digest for wards within the district is an example of this type. The second is the area classification, in which a selection of variables are analysed using some form of multivariate statistical procedure. Such classifications were first produced on a nationwide basis in the 1970s (Webber and Craig, 1977) and have been replicated both nationally and locally in the 1980s (Cloke and Edwards, 1986). As indicated, the use of geographical information systems will enable further forms of data integration to become more commonplace. However, the problem of accessing suitable data sources will remain as an important issue, especially for those researchers who need information that contains social rather than simply population elements.

Conclusion

The growth of interest in social change in the countryside has broadened the scope of rural geography. Those geographers who would claim this disciplinary allegiance could be studying such diverse topics as housing, transport, employment, recreation and tourism, agriculture, land use, social welfare and education. This spread of interests clearly presents difficulties for any review of source material to reflect the full scope of the discipline. The five types of local statistical source that have been looked at are inevitably a selection from the whole range and many local authorities and voluntary organizations will hold other kinds of data that would provide useful information for researching rural social change. For example, certain counties have been particularly aware of the problems faced by their increasing elderly population and have conducted surveys to identify whether the needs of this group of people in the countryside are being satisfied. Some of the sources that have been discussed are well known and generally used by researchers in the field; others, perhaps in particular the local ones, are slightly less familiar.

A difficulty with certain of the national sources identified is that, apart from the Population Census and the Vital Statistics, these data tend not to be available at a sufficiently fine level of geographical detail to enable analysis of social change in specific areas. The local sources, especially where there is comprehensive coverage that is capable of being integrated geographically for a particular local authority, are very useful for examining social change. The problem with such sources is that this level of coverage is not available across the whole country and even where present there may be difficulties for researchers to gain access to the data, especially in a computerized form.

It is tempting to feel that a combination of local and national data from these sources would help in the study of rural population and social change. However, the problems indicated in attempting to assemble a disparate range of sources into a coherent whole should not be under-estimated. A general-purpose survey that had sufficient spatial depth as well as measurement of the social and economic characteristics of the rural population would go some considerable way towards solving the problem.

REFERENCES

Acock, A. C. and Deseran, F. A. (1986) Off-farm employment by women and marital instability, *Rural Sociology*, Vol. 5, no. 3, pp. 314–27.

ACORA (1990) *Faith in the Countryside*, Churchman & ACORA, Worthing and London.

Agricultural Training Board (1985) *Review of Training in Agriculture and Horticulture, 1984–5*, Beckenham, Kent.

Alderson, M. (1988) Demographic and health trends in the elderly, in N. Wells and C. Freer (eds.), op. cit.

Ambrose, P. (1974) *The Quiet Revolution*, Chatto & Windus, London.

Ascher, K. (1987) *The Politics of Privatisation: Contracting Out Public Services*, Macmillan, London.

Association of District Councils (1986) *The Rural Economy at the Crossroads*, London.

Bailey, J. M. and Layzell, A. D. (1983) *Special Transport Service for Elderly and Disabled People*, Gower, London.

Bain, G. S. and Price, R. (1983) Union growth: dimensions, determinants and destiny in Britain, in G. S. Bain (ed.) *Industrial Relations in Britain*, Blackwell, Oxford.

Ball, R. M. (1987) Intermittent labour forms in U.K. agriculture: some implications for rural areas, *Journal of Rural Studies*, Vol. 3, no. 2, pp. 133–50.

Bamberger, M. (1988) *The Role of Community Participation in Development Planning and Project Management*, World Bank, Washington, DC.

Banister, D. and Norton, F. (1988) The role of the voluntary sector in the provision of rural services – the case of transport, *Journal of Rural Studies*, Vol. 4, no. 1, pp. 57–72.

Barlow, J. and Savage, M. (1986) The politics of growth: cleavage and conflict in a Tory heartland, *Capital and Class*, Vol. 30, pp. 156–82.

Barsh, R. and Gale, J. (1982) US economic development policy: the urban–rural dimension, in G. Stevens-Redburn and T. Buss (eds.) *Public Policies for Distressed Communities*, D. C. Heath, Lexington, Mass.

Bebbington, A. C. (1979) Changes in the provision of social services to the elderly in the community over fourteen years, *Social Policy and Administration*, Vol. 13, no. 2, pp. 111–23.

Beechey, V. and Perkins, T. (1987) *A Matter of Hours: Women, Part-Time Work and the Labour Market*, Polity Press, Cambridge.

Bell, P. and Cloke, P. (1988) *Bus Deregulation in the Powys/Clywd Study Area: The Final*

Report (TRRL Contractor Report 104), Transport and Road Research Laboratory, Crowthorne.

Bell, P. and Cloke, P. (1989) The changing relationship between the private and public sectors: privatisation and rural Britain, *Journal of Rural Studies*, Vol. 5, no. 1, pp. 1–15.

Bell, P. and Cloke, P. (1990) (eds.) *Deregulation and Transport*, Fulton, London.

Bell, W. (1958) Social choice, lifestyle and suburban residence, in W. Dobriner (ed.) *The Suburban Community*, Putnam, New York, NY.

Bennett, S. (1977) Housing need and the housing market, in G. Williams (ed.) *Community Development in Countryside Planning*, Department of Town and Country Planning, University of Manchester.

Bentham, G. and Haynes, R. (1984) Health, personal mobility and the use of health services in rural Norfolk, *Journal of Rural Studies*, Vol. 1, pp. 231–40.

Blacksell, M., Economides, K. and Watkins, C. (1991, forthcoming) *Justice Outside the City: Access to Legal Services in Rural Britain*, Longman, London.

Bolton, N. and Chalkley, B. (1990) The rural population turnaround: a case-study of North Devon; *Journal of Rural Studies*, Vol. 6, no. 1, pp. 29–43.

Bonnar, D. (1979) Migration in the south-east of England: an analysis of the inter-relationships of housing, socio-economic status and housing demand, *Regional Studies*, Vol. 13, pp. 345–59.

Bouquet, M. (1985) *Family, Servants and Visitors: The Farm Household in Nineteenth and Twentieth Century Devon*, Geo Books, Norwich.

Bowler, I. (1985) Some consequences of the industrialization of agriculture in the European Community, in M. J. Healey and B. W. Ilbery (eds.), op. cit.

Bowler, I. and Lewis, G. (1987) The decline of private rented housing in rural areas: a case study of villages in Northamptonshire, in D. G. Lockhart and B. Ilbery (eds.) *The Future of the British Rural Landscape*, Geobooks, Norwich.

Bracey, H. E. (1952) *Social Provision in Rural Wiltshire*, Methuen, London.

Bracey, H. E. (1958) A note on rural depopulation and social provision, *Sociological Review*, Vol. 6, pp. 67–74.

Bracey, H. E. (1959) *English Rural Life*, Routledge, London.

Bracey, H. E. (1970) *People and the Countryside*, Routledge, London.

Bradley, T. (1987) Poverty and dependency in rural England, in P. Lowe, T. Bradley and S. Wright (eds.) op. cit.

Brotherton, I. (1981) *Conflict, Consensus, Concern and the Administration of Britain's National Parks*, Department of Landscape Architecture, University of Sheffield.

Brown, J. and Ward, S. (1990) *The Village Shop*, David & Charles, Newton Abbot.

Brown, L. and Moore, E. (1970) The intra-urban migration process: a perspective, *Geografiska Annaler*, Vol. 52B, pp. 1–13.

Buchanan, J. (1979) Physically handicapped in a rural environment, *Housing and Planning Review*, Vol. 35, no. 4, pp. 8–11.

Buchanan, J. (1983) *The Mobility of Disabled People in a Rural Environment*, Royal Association for Disability and Rehabilitation, London.

Buchanan, W. I., Errington, A. J. and Giles, A. K. (1982) *The Farmer's wife: Her Role in the Management of the Business* (Study no. 2), Farm Management Unit, University of Reading.

Buller, H. and Wright, S. (eds.) (1990) *Rural Development: Problems and Practices*, Avebury, Aldershot.

Burrell, A., Hill, B. and Midland, J. (1984) *Statistical Handbook of UK Agriculture*, Macmillan, Basingstoke.

Byron, R. (ed.) (1988) *Public Policy and the Periphery: Problems and Prospects in Marginal Regions*, Proceedings of the Ninth International Seminar on Marginal Regions, Skye and Lewis, Scotland, July 1987, International Society for the Study of Marginal Regions, Halifax, Nova Scotia.

Cadwallader, M. (1986) Migration and intra-urban mobility, in M. Pacione (ed.) *Popula-*

tion Geography: Progress and Prospect, Croom Helm, London.

Campbell, H. F. (1920) *Highland Reconstruction*, Alex Maclaren, Glasgow.

Capstick, M. (1987) *Housing Dilemmas in the Lake District*, Centre for North West Regional Studies, University of Lancaster.

Carr, J. D. (ed.) (1986) *Passenger Transport: Planning for Radical Change*, Gower, Aldershot.

Centre of Management in Agriculture (1987) *Farming Facts* (2nd edn.), Corby.

Champion, A. G. (1981) Population trends in rural Britain, *Population Trends*, Vol. 26, pp. 20–3.

Champion, A. G. (1982) Rural-urban contrasts in Britain's population change 1961–81, in A. Findlay (ed.) *Recent National Population Change*, Institute of British Geographers, Durham University.

Champion, A. G. (1989a) *Counterurbanization: The Changing Pace and Nature of Population Deconcentration*, Edward Arnold, London.

Champion, A. G. (1989b) United Kingdom: population deconcentration as a cyclic phenomenon, in A. G. Champion (1989a), op. cit.

Champion, A. G., Green, A. E., Owen, D. W., Ellis, D. J. and Coombes, M. G. (1987) *Changing Places. Britain's Demographic, Economic and Social Complexion*, Edward Arnold, London.

Champion, A. G. and Townsend, A. R. (1990a) *Contemporary Britain*, Edward Arnold, Sevenoaks.

Champion, A. G. and Townsend, A. R. (1990b) Demographic forces and the reshaping of rural England in the late twentieth century, in ACORA, op. cit., Appendix D.

Chisholm, M. (1984) The Development Commission's factory programme, *Regional Studies*, Vol. 18, pp. 514–17.

Chisholm, M. (1985) The Development Commission's employment programmes in rural England, in M. J. Healey and B. W. Ilbery (eds.), op. cit.

Clark, A. M. and Thomas, F. G. (1990) The geography of the 1991 Census, *Population Trends*, Vol. 60, pp. 9–15.

Clark, D. (1990) *Affordable Housing in Villages* (report to the RDC), ACRE, Cirencester.

Clark, D. and Woollett, S. (1990) *English Village Services in the Eighties* (Rural Research Series no. 7), Rural Development Commission, London.

Clark, G. (1979) Farm amalgamation in Scotland, *Scottish Geographical Magazine*, Vol. 95, pp. 93–107.

Clark, G. (1982a) Developments in rural geography, *Area*, Vol. 14, pp. 249–54.

Clark, G. (1982b) *Housing and Planning in the Countryside*, Wiley, Chichester.

Clark, G. (1982c) *The Agricultural Census – United Kingdom and United States* (CATMOG 35), Geo Books, Norwich.

Clark, G., Johnson, J. H. and McAuley, A. (1989) Farm managers' salaries, *Farm Management*, Vol. 7, pp. 119–26.

Clark, W. and Onaka, J. (1983) Life cycle and housing adjustment as explanations of residential mobility, *Urban Studies*, Vol. 20, pp. 47–57.

Clarkson, S. (1980) *Jobs in the Countryside – Some Aspects of the Work of the Rural Industries Bureau and the Council for Small Industries in Rural Areas, 1910–1979* (Occasional Paper 2), Department of Environmental Science and Countryside Planning, Wye College.

Cloke, P. (1977) An index of rurality for England and Wales, *Regional Studies*, Vol. 11, pp. 313–46.

Cloke, P. (1979) *Key Settlements in Rural Areas*, Methuen, London.

Cloke, P. (1980) New emphases for applied rural geography, *Progress in Human Geography*, Vol. 4 pp. 181–217.

Cloke, P. (1983a) *An Introduction to Rural Settlement Planning*, Methuen, London.

Cloke, P. (1983b) Counter-urbanisation – a rural perspective, *Geography*, Vol. 70, pp. 13–23.

Cloke, P. (1990) Political economy approaches and a changing rural geography, *Rural History*, Vol. 1, pp. 123–9.

Cloke, P. and Edwards, G. (1986) Rurality in England and Wales 1981: a replication of the 1971 index, *Regional Studies*, Vol. 20, no. 4, pp. 289–306.

Cloke, P. and Little, J. (1987) The impact of decision making on rural communities: an example from Gloucestershire, *Applied Geography*, Vol. 7, pp. 55–77.

Cloke, P. and Little, J. (1990) *The Local State?*, Oxford University Press.

Cloke, P. and Moseley, M. (1990) Rural geography in Britain, in P. Lowe and M. Bodiguel (eds.), op. cit.

Cloke, P. and Thrift, N. (1987) Intra-class conflict in rural areas, *Journal of Rural Studies*, Vol. 3, pp. 321–33.

Cloke, P. and Thrift, N. (1990) Class and change in rural Britain in T. Marsden, P. Lowe and S. Whatmore (eds.) *Rural Restructuring: Global Processes and Local Responses*, Fulton, London.

Clout, H. D. (1972) *Rural Geography: An Introductory Survey*, Pergamon Press, Oxford.

Clout, H. (1984) *A Rural Policy for the EEC*, Methuen, London.

Coffey, W. and Polese, M. (1985) Local development: conceptual bases and policy implications, *Regional Studies*, Vol. 19, pp. 85–93.

Collins, I. and Little, J. (1989) *Women and Employment in Rural Hampshire* (unpublished report to the Hampshire Rural Development Commission), Winchester

Constable, M. (1988) Speech to HCT seminar, London.

Cooke, P. and Hulme, D. (1988) The compatibility of market liberalisation and local economic development strategies, *Regional Studies*, Vol. 22, pp. 221–32.

Cottam, M. B. and Knox, P. L. (1982) *The Highlands and Islands: A Social Profile* (Occasional Paper no. 6), Department of Geography, University of Dundee.

Craig, J. (1988) Local authority urban-rural indicators compared, *Population Trends*, Vol. 51, pp. 30–8.

Cullingford, D. and Openshaw, S. (1982) Identifying areas of rural deprivation using social area analysis, *Regional Studies*, Vol. 16, pp. 409–17.

Cumbria Countryside Conference (1979) *Rural Housing* (Report).

Cumbria County Council (1987) *Joint Rural Development Programme for Cumbria (1988–1991)*, Kendal.

Curtis, L. P. (1963) *Coercion and Conciliation in Ireland 1880–1892: A Study in Conservation Unionism*, Princeton University Press, NJ.

Davies, D., Pack, C., Seymour, S., Short, C., Watkins, C. and Winter, M. (1990a) *Rural Church Project: The Views of Rural Parishioners* (Occasional Paper no. 14), Centre for Rural Studies, Cirencester.

Davies, D., Pack, C., Seymour, S., Short, C., Watkins, C. and Winter, M. (1990b) *Parish Life and Rural Religion* (Occasional Paper no. 13), Centre for Rural Studies, Cirencester.

Davies, D., Pack, C., Seymour, S., Short, C., Watkins, C. and Winter, M. (1990c) *The Clergy Life* (Occasional Paper no. 12), Centre for Rural Studies, Cirencester.

Davison, J. (1985) *Integrated Rural Development in Britain: Retrospect and Prospect* (Graduate Discussion Paper 10), Department of Geography, University of Durham.

Dawson, J. A. and Kirby, D. A. (1979) *Small Scale Retailing in the UK*, Saxon House, Farnborough.

Day, G., Rees, G. and Murdoch, J. (1989) Social Change, rural localities, and the state: the restructing of rural Wales, *Journal of Rural Studies*, Vol. 5, pp. 227–44.

Dean, K. G., Shaw, D. P., Brown, B. J. H., Perry, R. W. and Thorneycroft, W. T. (1984) Counterurbanization and the characteristics of persons migrating to West Cornwall, *Geoforum*, Vol. 15, pp. 177–90.

Deavers, K. L., Hoppe, R. A. and Ross, P. J. (1986) Public policy and rural poverty: a view from the 1980s, *Policy Studies Journal*, Vol. 15, pp. 291–309.

Department of Transport (with Scottish Office and Welsh Office) (1984) *Buses* (Cmnd 9300, white paper), HMSO, London.

Development Commission (1984a) *Guidelines for Development of Rural Development Programmes*, London.

Development Commission (1984b) *The Designation of Rural Development Areas*, London.

Development Commission (1987) *Opportunity Through Diversity*, London.

Dobbs, A. C. (1989) Telecottages in the United Kingdom, in C. Watkins (ed.), op. cit.

DoE (1983) *Modifications to the Cumbria and Lake District Joint Structure Plan*.

DoE (1988a) *Housing in Rural Areas* (a statement by the Secretary of State for the Environment), HMSO, London.

DoE (1988b) *Housing in Rural Areas: Village Housing and New Villages* (discussion paper), HMSO, London.

Donnelly, P. and Harper, S. (1987) British rural settlements in the hinterlands of conurbations: a classification, *Geografiska Annaler*, Vol. 69B, pp. 55–63.

Donnison, D. (1984) The progressive potential of privatisation, in J. Le Grand and R. Robinson (eds.), op. cit.

Douglas, D. (1987) Community development: scope, issues, potentials, *Journal of Community Development*, Vol. 1, pp. 17–23.

Drudy, P. J. (1978) Depopulation in a prosperous agricultural sub-region, *Regional Studies*, Vol. 12, pp. 49–60.

Drudy, P. J. and Drudy, S. M. (1979) Population mobility and labour supply in rural regions in Norfolk and the Galway Gaeltacht, *Regional Studies*, Vol. 13, pp. 92–9.

Dunbar, C. S. (1984) Hereford and Worcester: a model for the future?, *Buses*, Vol. 36, no. 355, pp. 444–6.

Dunleavy, P. and O'Leary, D. B. (1987) *Theories of the State: The Politics of Liberal Democracy*, Macmillan, Basingstoke.

Dunn, M., Rawson, M. and Rogers, A. W. (1981) *Rural Housing, Competition and Choice*, Allen & Unwin, London.

Dunn, M. and Swindell, K. (1972) Electoral registers and rural migration: a case study of Herefordshire, *Area*, Vol. 4, pp. 39–42.

Economic and Social Committee (1989) *Opinion on the Future of Rural Society* (European Communities Economies and Social Committee, CES(89) 1026), Brussels.

Employment Gazette (1988) Department of Employment, London, April.

England, J. R., Hudson, K. I., Masters, R. J., Powell, R. S. and Shortridge, J. D. (1985) *Information Systems for Policy Planning and Local Government*, Longman, Harlow.

Errington, A. J. (1980) Occupational classification in British agriculture, *Journal of Agricultural Economics*, Vol. 31, pp. 73–81.

Errington, A. J. (1983) The farmer's wife: her role in the farm business, in R. B. Tranter (ed.) *Strategies for Family-Worked Farms in the UK* (Paper 15), Centre for Agricultural Strategy, Reading.

Errington, A. J. (1990) Investigating rural employment in England, *Journal of Rural Studies*, Vol. 6, pp. 67–84.

European Commission (1988) *The Future of Rural Society* (COM (88) 371, final), *Bulletin of the European Communities* (Supplement 4/88), Brussels.

Eurostat (1987) *Farm Structure 1985 Survey: Analysis of Results* (plus corrigendum n.d.), Statistical Office of the European Communities, Brussels.

Eurostat (1989) *Agriculture: Statistical Yearbook, 1989*, Statistical Office of the European Communities, Brussels.

Evans, A. (1988) Hereford: a case-study of bus deregulation, *Journal of Transport Economics and Policy*, Vol. 22, no. 3, pp. 283–306.

Family Policy Studies Centre (1988) *An Ageing Population*, London.

Farrington, J. H. (1986) Deregulation of the British bus system, *Geography*, Vol. 71, p. 258.

Fearn, R. (1987) Rural health care: a British success or tale of unmet need?, *Social Science and Medicine*, Vol. 24, pp. 263–74.

Fennell, J. (1988) Rural health needs ignored, *Rural Viewpoint*, Issue 24, p. 1.

Fernadez, R. and Dillman, D. (1979) The influence of community attachment on geographic mobility, *Rural Sociology*, Vol. 44, no. 2, pp. 345–60.

Fielding, A. (1986) Counterurbanisation, in M. Pacione (ed.) *Population Geography: Progress and Prospect*, Croom Helm, London.

Fielding, A. (1989) Inter-regional migration and social change: a study of south-east England based upon data from the Longitudinal Study, *Transactions of the Institute of British Geographers*, Vol. 14, pp. 24–36.

Fleming, P. (1979) *Villagers and Strangers*, Schenkman, Cambridge, Mass.

Ford, G. and Taylor, S. (1983) Risk groups and selective case finding in an elderly population, *Social Science and Medicine*, Vol. 17, no. 10, pp. 647–55.

Forrest, R. and Murie, A. (1988) *Selling the Welfare State: The Privatisation of Public Housing*, Routledge, London.

Fothergill, S. and Gudgin, G. (1979) Regional employment change: a sub-regional explanation, *Progress in Planning*, Vol. 12, pp. 155–219.

Fothergill, S., Gudgin, G., Kitson, M. and Monk, S. (1985) Rural industrialization: trends and cases, in M. J. Healey and B. W. Ilbery (eds.), op. cit.

Frankenberg, R. (1966) *Communities in Britain*, Penguin Books, Harmondsworth.

Frye, A. (1986) A new age of enlightenment, *Transport*, Vol. 7, no. 6, pp. 263–5.

Furness, G. W. (1983) The importance, distribution and net incomes of small farm businesses in the United Kingdom, in R. B. Tranter (ed.) *Strategies for Family-Worked Farms in the United Kingdom* (paper 15), Centre for Agricultural Strategy, Reading.

Gamble, A. (1988) *The Free Economy and the Strong State: The Politics of Thatcherism*, Macmillan, Basingstoke.

Gant, R. L. and Smith, J. A. (1985a) Cotswold buses – endangered species?, *Buses*, Vol. 37, pp. 58–61.

Gant, R. L. and Smith, J. A. (1985b) Cotswold people need the bus, *Buses*, Vol. 37, pp. 106–8.

Gant, R. L. and Smith, J. A. (1985c) Deregulation in the Cotswolds, in *Proceedings of the Twelfth Seminar on Rural Public Transport*, Polytechnic of Central London

Gant, R. L. and Smith, J. A. (1988) Journey patterns of the elderly and disabled in the Cotswolds: a spatial analysis, *Social Science and Medicine*, Vol. 27, no. 2, pp. 173–80.

Gaskell, P. (1968) *Morvern Transformed: A Highland Parish in the Nineteenth Century*, Cambridge University Press, London.

Gasson, R. (1966) *The Influence of Urbanisation on Farm Ownership and Practice: Some Aspects of the Effect of London on Farms and Farm People in Kent and Sussex* (Report no. 7), Studies in Rural Land Use, Wye College.

Gasson, R. (1981) The roles of women on farms: a pilot study, *Journal of Agricultural Economics*, Vol. 32, pp. 11–20.

Gasson, R. (1983a) *Gainful Occupations of Farm Families*, School of Rural Economics, Wye College.

Gasson, R. (1983b) Part-time farming: a strategy for family farms?, in R. B. Tranter (ed.) *Strategies for Family-Worked Farms in the UK* (paper 15), Centre for Agricultural Strategy, Reading.

Gasson, R. (1989) *Farm Work by Farmer's Wives*, Farm Business Unit, Department of Agricultural Economics, Wye College.

Gilg, A. W. (1985) *An Introduction to Rural Geography*, Edward Arnold, London.

Glaister, S. and Mulley, C. (1983) *Public Control of the British Bus Industry*, Gower, Aldershot.

Gloucestershire County Council (1985a) *Rural Development Programme*, Gloucester.

Gloucestershire County Council (1985b) *Voluntary Transport Schemes in Gloucestershire*, Gloucester.

Gloucestershire County Council (1986a) *Gloucestershire Structure Plan*, Gloucester.

Gloucestershire County Council (1986b) *North Cotswolds Surveys. Parish Profiles*,

Gloucestershire County Social Services, Gloucester.

Goddard, G. (1988) Occupational accident statistics 1981–5, *Employment Gazette*, January, pp. 15–21.

Gordon, P. (1979) Deconcentration without a 'clean break', *Environment and Planning A*, Vol. 11, pp. 281–9.

Grafton, D. J. (1982) Net migration, outmigration and remote rural areas, *Area*, Vol. 14, no. 4, pp. 313–18.

Gray, M. (1957) *The Highland Economy 1750–1850*, Oliver & Boyd, Edinburgh.

Green, C. (1986) Rural Development Areas – progress and problems, *Planner*, Vol. 72, pp. 18–19.

Green, D. G. (1987) *The New Right: The Counter Revolution in Political, Economic and Social Thought*, Wheatsheaf, Brighton.

Greenwood, J. (1989) *Planning for Low-Cost Rural Housing* (Working Paper no. 112), School of Planning, Oxford Polytechnic.

Grieco, M. (1990) Development in the developed world: revealing the hidden agenda, in H. Buller and S. Wright (eds.) op. cit.

Grundy, E. (1987) Retirement migration and its consequences in England and Wales, *Ageing and Society*, Vol. 7, pp. 57–82.

Guardian (1989) White paper charts switch to private care, 17 November.

Halford, S. (1987) *Women's Initiatives in Local Government; Tokenism or Power* (Working Paper 58), Department of Urban and Regional Studies, University of Sussex.

Hall, P., Thomas, R., Gracey, H. and Drewett, R. (1973) *The Containment of Urban England*, Allen & Unwin, London.

Hamnett, C. and Randolph, W. (1983) The changing population distribution of England and Wales, 1961–81: clean break or consistent progression?, *Built Environment*, Vol. 8, pp. 272–80.

Hampshire County Council (1966) *South Hampshire Study*, Winchester.

Hampshire County Council (1980) *Mid-Hampshire Structure Plan: Approved Written Statement*, Winchester.

Hansen, J. C. (1988) Official policies in marginal regions: temporary relief or new deal?, in R. Byron (ed.), op. cit.

Harper, S. (1987a) A humanistic approach to the study of rural populations, *Journal of Rural Studies*, Vol. 3, pp. 309–19.

Harper, S. (1987b) Rural reference groups and images of place, in D. Pocock (ed.) *Humanistic Approaches in Geography* (Occasional Paper 23), Department of Geography, University of Durham.

Harper, S. (1987c) The kinship network of the rural aged: a comparison of the indigenous elderly and the retired inmigrant, *Ageing and Society*, Vol. 7, pp. 303–27.

Harper, S. (1987d) The rural-urban interface in England: a framework of analysis, *Transactions of the Institute of British Geographers*, Vol. 12, pp. 284–302.

Harris, C. (1974) *Hennage. A Social System in Miniature*, Holt, Rinehart & Winston, London.

Hastings, M. (1987–8) Farming wives as business managers, *Farm Management*, Vol. 6, pp. 309–15.

Haynes, R. (1987) *The Geography of Health Services in Britain*, Croom Helm, London.

Heald, D. A. (1984) Privatisation: analysing its appeal and limitations, *Fiscal Studies*, Vol. 5, no. 1, pp. 36–46.

Heald, D. A. and Thomas, D. (1986) Privatization as theology, *Public Policy and Administration*, Vol. 1, no. 2, pp. 49–66.

Healey, M. J. and Ilbery, B. W. (eds.) (1985) *The Industrialization of the Countryside*, Geo Books, Norwich.

Higginson, M. (1984) *Buses after 'Buses': Possible Implications of the 1984 White Paper*, Department of Extra-Mural Studies, University of London.

Hill, B. (1984) Information on farmers' incomes: data from Inland Revenue sources,

Journl of Agricultural Economics, Vol. 35, pp. 39–50.

Hillier, J. (1982) The role of CoSIRA factories in the provision of employment in rural Eastern England, in M. Moseley (ed.) *Power, Planning and People in Rural East Anglia*, Geo Books, Norwich.

Hillman, M. and Whalley, A. (1979) *Walking in Transport*, Policy Studies Institute, London.

Hirsch, F. (1978) *Social Limits to Growth*, Routledge, London.

Hitchcock, A. (1980) Mobility and the elderly, in ECMT *Report of the Fifty-First Round Table of Transport Economics*, Paris.

HMSO (1989) *Caring for People – Community Care in the Next Decade and Beyond*, London.

Hodge, I. (1986) The scope and context of rural development, *European Review of Agricultural Economics*, Vol. 13, pp. 271–82.

Hodge, I. and Whitby, M. (1979) *New Jobs in the Eastern Borders: An Economic Evaluation of the Development Commission's Factory Building Programme* (Monograph 8), Agricultural Adjustment Unit, University of Newcastle upon Tyne.

Hoggart, K. (1988) Not a definition of rural, *Area*, Vol. 20, pp. 35–40.

Hoggart, K. and Buller, H. (1987) *Rural development – A Geographical Perspective*, Croom Helm, London.

Holdcroft, L. E. and Jones, G. E. (1982) The rise and fall of community development in developing countries, 1950–65: a critical analysis and implication, in G. E. Jones and M. Rolls (eds.) *Progress in Rural Extension and Community Development*, Wiley, Chichester.

House of Lords (1990) *The Future of Rural Society* (House of Lords Session 1989–90, 24th Report, Select Committee on the European Communities, HL Paper 80-I), HMSO, London.

Howarth, R. W. (1985) *Farming for Farmers? A Critique of Agricultural Support Policy* (Hobart Paperback no. 20), Institute of Economic Affairs, London.

Johnson, J. H. (1967) Harvest migration from nineteenth-century Ireland, *Transactions of the Institute of British Geographers*, Vol. 41, pp. 97–112.

Johnson, M (1988) A throng at twilight, *Guardian*, 6 December.

Johnstone, W. D., Nicholson, C, Stone, M. K. and Taylor, R. E. (1990) *Countrywork*, ACRE, Cirencester.

Joint Unit for Research in the Urban Environment (1983) *An Evaluation of Development Commission Activities in Selected Areas – Summary Report*, The Unit, University of Aston.

Jones, C. and Armitage, B. (1990) Population change within area types: England and Wales, 1971–88, *Population Trends*, Vol. 60, pp. 25–32.

Jones, H. (1981) *A Population Geography*, Harper & Row, London.

Jones, H. (1990) *Population Geography*, Paul Chapman, London.

Jones, H. *et al.* (1986) Peripheral counter-urbanisation: findings from an integration of Census and survey data in Northern Scotland, *Regional Studies*, Vol. 20, no. 1, pp. 15–26.

Jones, V. (1986) Trial areas – the lessons of experience, in J. D. Carr (ed.), op. cit.

Kilvington, R. P. and Cross, A. K. (1986) *Deregulation of Express Coach Services in Britain*, Gower, Aldershot.

King, D. S. (1987) *The New Right: Politics, Markets and Citizenship*, Macmillan, Basingtoke.

Kontuly, T., Wiard, S. and Vogelsang, R. (1986) Counterurbanistion in the Federal Republic of Germany, *Professional Geographer*, Vol. 38, pp. 170–81.

Law, D. and Howes, K. (1972) *Mid-Wales – An Assessment of the Development commission's Factory Programme*, HMSO, London.

Lawton, R. (1968) Population changes in England and Wales in the late nineteenth century: an analysis of trends by registration districts, *Transactions, Institute of British Geographers*, Vol. 44, pp. 55–74.

Lawton, R. (1977) Regional population trends in England and Wales, 1750–1971, in J. Hobcraft and P. Rees (eds.) *Regional Demographic Development*, Croom Helm, London.

LDSPB (1977a) *Draft National Park Plan*, Kendal.

LDSPB (1977b) Minutes, Planning Committee Kendal, 2 November.

LDSPB (1978) *National Park Plan*, Kendal.

LDSPB (1979) *27th Annual Report – 1978/79*, Kendal.

LDSPB (1980) *Housing in the Lake District National Park* (Supplementary Statement no. 1 for the examination in public of the Cumbria and Lake District Joint Structure Plan), Kendal.

LDSPB (1981) *29th Annual Report, 1980/81*, Kendal.

LDSPB (1982) *30th Annual Report, 1981/82*, Kendal.

LDSPB (1988) *A Lake District Protection Bill* (Report by National Park Officer to the Planning Policy Committee), Kendal, 15 December.

Lee, C. H. (1979) *British Regional Employment Statistics*, Cambridge University Press.

Le Grand, J. (1983) Is privatisation always such a bad thing?, *New Society*, Vol. 64, no. 1064, pp. 7–9, 7 April.

Le Grand, J. and Robinson, R. (1984) (eds.) *Privatisation and the Welfare State*, Allen & Unwin, London.

Leominster Marches Project (1986) *Rural Community Development in Action*, Community Projects Foundation, London.

Leslie, G. and Richardson, A. (1961) Life cycle, career pattern and the decision to move, *American Sociological Review*, Vol. 26, pp. 894–902.

Lewis, G. J. (1982) *Human Migration*, Croom Helm, London.

Lewis, G. J. (1989) Counterurbanization and social change in the rural South Midlands, *East Midland Geographer*, Vol. 11, pp. 3–12.

Little, J. (1986) Social class and planning policy: a study of two Wiltshire villages, in P. Lowe, A. J. Bradley and S. Wright (eds.), op. cit.

Little, J. (1987a) Gender relations in rural areas; the importance of women's domestic role, *Journal of Rural Studies*, Vol. 3, no. 4, pp. 335–42.

Little, J. (1987b) Woman's 'place' in the rural community. Unpublished paper to IBG Rural Geography Study Group, Institute of British Geographers Annual Conference, Portsmouth.

Little, J. (1990) Women's employment in the food system, in T. Marsden and J. Little (eds.) *Critical Perspectives on the Food System*, Gower, Aldershot.

Little, J. and Ross, K. (1990) women and Employment in Rural Areas (Report to the Rural Development Commission), RDC, London.

Long, L. (1988) *Migration and Residential Mobility in the United States*, Russel Sage, New York, NY.

Loughlin, M. (1984) *Local Needs Policies and Development Control Strategies* (SAUS Working Paper 42, University of Bristol.

Lowe, P. and Bodiguel, M. (eds.) (1990) *Rural Studies in Britain and France*, Belhaven, London.

Lowe, P., Bradley, T. and Wright, S. (eds.) (1986) *Deprivation and Welfare in Rural Areas*, Geo Books, Norwich.

Lundqvist, L. J. (1988) Privatization: towards a concept for comparative policy analysis, *Journal of Public Policy*, Vol. 8, no 1, pp. 1–19.

Macfarlane, G. (1981) Shetlanders and incomers: change, conflict and emphasis in social perspectives, in L. Holy and M. Stuchtek (eds.) *The Structure of Folk Models* (ASA Monograph 20), Academic Press, London.

Mackay, G. A. (1973) Regional planning problems: Scotland, in M. Broady (ed.) *Marginal Regions: Essays in Social Planning*, National Council of Social Service, London.

Mackay, G. A. and Laing, G. (1982) *Consumer Problems in Rural Areas*, Scottish Consumer Council, Glasgow.

Magnusson, M. (1968) Highland administration, in D. S. Thomson and I. Grimble (eds.) *The Future of the Highlands*, Routledge & Kegan Paul, London.

Marks, H. F. (1989) *A hundred Years of British Food and Farming: A Statistical Survey* (ed. A. K. Britton) Taylor & Francis, London.

Martin, P. C. (1977) The age structure of Scottish farm workers, *Scottish Agricultural Economics*, Vol. 27, pp. 87–91.

Mayo, J. (1983) Transport needs of the elderly, in T. Aldous (ed.) *We All Need the Bus* (Convention Report), Bus and Coach Council, London.

McCleery, A. (1984) *The Role of the Highland Development Agency: With Particular Reference to the Work of the Congested Districts Board 1897–1912* (unpublished PhD thesis), University of Glasgow.

McCleery, A. (1988) The Highland Board reviewed: a note on the analysis of economic and social change, *Scottish Geographical Magazine*, Vol. 104, no. 3, pp. 171–5.

McCleery, A. *et al.* (1987) *Review of the Highlands and Islands Development Board: Economic and Social Change in the Highlands and Islands* (Economics and Statistics Unit Research Paper no. 13), Industry Department for Scotland, Edinburgh.

McDermott, K. and Dench, S. (1983) *Youth Opportunity Programmes in Rural Areas*, Manpower Services Commission, Sheffield.

McIntosh, F. (1972) A survey of workers leaving Scottish farms, *Scottish Agricultural Economics*, Vol. 22, pp. 147–52.

McIntosh, M. (1978) The State and the oppression of women, in A. Kuhn and A. Wolpe (eds.) *Feminism and Materialism*, Routledge & Kegan Paul, London.

McLaughlin, B. (1986) The rhetoric and the reality of rural deprivation, *Journal of Rural Studies*, Vol. 2, no. 4, pp. 291–307.

McLaughlin, B. (1987) Rural policy into the 1990s: self help or self-deception, *Journal of Rural Studies*, Vol. 3, no. 4, pp. 361–4.

McNab, A. (1984) *Integrated Rural Development in Britain: Concepts and Practices* (Gloucestershire Papers in Local and Regional Planning no. 22), Gloucestershire College of Arts and Technology, Gloucester.

Michelson, W. (1977) *Environmental Choice, Human Behaviour and Residential Satisfaction*, Oxford University Press, New York, NY.

Minay, C. (1985) *The Development Commission's Rural Industrial Development Programme: A Review of Progress 1945–85* (Working Paper 87), Department of Town Planning, Oxford Polytechnic.

Ministry of Agriculture and Fisheries (1946) *National Farm Survey of England and Wales (1941–43): A Summary Report*, HMSO, London.

Ministry of Agriculture, Fisheries and Food (1967) *The Changing Structure of the Farm Labour Force in England and Wales: 1946–65*, London.

Ministry of Agriculture, Fisheries and Food (1987) *Farm Incomes in the United Kingdom: 1987*, HMSO, London.

Ministry of Agriculture, Fisheries and Food (1988) *Agricultural Labour in England and Wales: Earnings, Hours and Numbers of Persons*, London.

Ministry of Agriculture, Fisheries and Food (1989) *Agricultural and Horticultural Returns – Final Results of the June 1988 Census in England and Wales*, Agricultural Census Branch, Guildford.

Ministry of Agriculture, Fisheries and Food, Department of Agriculture and Fisheries for Scotland (1968) *A Century of Agricultural Statistics: Great Britain 1866–1966*, HMSO, London.

Ministry of Agriculture, Fisheries and Food, Department of Agriculture and Fisheries for Scotland, Department of Agriculture for Northern Ireland, Welsh Office (1989) *Agricultural Statistics, United Kingdom 1987*, HMSO, London.

Ministry of Housing and Local Government (1964) *The South East Study 1961–1981* HMSO, London.

Mohan, S. (1989) Continuity and change in state economic intervention, in A. Cochrane and J. Anderson (eds.) *Restructuring Britain, Politics in Transition*, Sage, London.

Morrison, P. A. and Wheeler, J. P. (1976) Rural renaissance in America?, *Population Bulletin*, Vol. 31, no. 3, pp. 1–27.

Moseley, M. J. (1979) *Accessibility: The Rural Challenge*, Methuen, London.

Moseley, M. J. (1980a) Is rural deprivation really rural?, *The Planner*, Vol. 66, p. 97.

Moseley, M. J. (1980b) *Rural Development and its Relevance to the Inner City Debate* (Inner City in Conflict, Paper 9), SSRC, London.

Moseley, M. J. (1985) *The Waveney Project: The Role of the Catalyst in Rural Community Development* (Social Works Monographs), University of East Anglia, Norwich.

Moseley, M. J., Harman, R. G., Coles, O. B. and Spencer, M. B. (1977) *Rural Transport and Accessibility*, Centre of East Anglian Studies, Norwich.

Moseley, M. J. and Packham, J. (1985) The distribution of fixed, mobile and delivery services in rural Britain, *Journal of Rural Studies*, Vol. 1, pp. 87–95.

Mulley, C. (1983) The background to bus regulation in the 1930 Road Traffic Act: economic, political and personal influences in the 1920s, *Journal of Transport History* (New Series), Vol. 4, no. 2, pp. 1–19.

Munton, R., Marsden, T. and Whatmore, S. (1990) Agricultural restructuring: current trends and prospects, in ACORA, op. cit., Appendix E.

Murphy, P. (1979) Migration of the elderly: a review, *Town Planning Review*, Vol. 50, pp. 84–93.

Murray, M. and Hart, M. (1989) Integrated rural development in Northern Ireland: a case study of local initiatives within County Tyrone, in L. Albrechts (ed.) *Regional Policy at the Cross Roads: European Perspectives*, Jessica Kingsley, London.

NAC Rural Trust (1987) *Village Homes for Village People*, London.

Newby, H. (1977) *The Deferential Worker: A Study of Farm Workers in East Anglia*, Allen Lane, London.

Newby, H. (1979) *Green and Pleasant Land?*, Hutchinson, London.

Newby, H. (1988) *The Countryside in Question*, Hutchinson, London.

Newby, H. (1990, forthcoming) Revitalising the countryside: the opportunities and pitfalls of counter-urban trends, *Royal Society of Arts, Journal*.

Newby, H., Bell, C., Rose, D. and Saunders, P. (1978) *Property Paternalism and Power: Class and Control in Rural England*, Hutchinson, London.

Northamptonshire County Council (1987) *East Northamptonshire and Huntingdonshire Rural Development Programme*, Northampton.

North Cotswold Voluntary Help Centre (1984–8) *Annual Reports*, Moreton-in-Marsh.

Northfield Report (1979) *Report of the Committee of Inquiry into the Acquisition and Occupancy of Agricultural Land (Cmnd 7599), HMSO, London.*

North Yorkshire County Council (1985) *Joint Rural Development Programme for North Yorkshire*, Northallerton.

Nutley, S. D. (1988) Unconventional modes of transport in rural Britain: progress to 1985, *Journal of Rural Studies*, Vol. 4, no. 1, pp. 73–86.

Nutley, S. D. (1989) *Unconventional and Community Transport in the UK*, Gordon & Breach Science Publishers, London.

O'Cinneide, M. (1988) Commentary on the effectiveness of official policies, in R. Byron (ed.), op. cit.

OPCS (1981a) *Census 1981: Definitions, Great Britain* (CEN 81 DEF), HMSO, London.

OPCS (1981b) *1981 Census, Preliminary Report, England and Wales*, HMSO, London.

OPCS (1988a) *Key Population and Vital Statistics, 1986* (VS no. 13/PP1, no. 9), HMSO, London.

OPCS (1988b) *The Prevalence of Disability Among Adults: OPCS Surveys of Disability in Great Britain* (Report 1), HMSO, London.

OPCS (1989) *Local Statistics, Small Area Statistics (Revised Proposals)*, London, July.

OPCS (1990) *Local Statistics, Small Area Statistics (Further Revisions)*, London, January.

Orwin, C. S. (ed.) (1944) *Country Planning: A Study of Rural Problems*, Oxford University Press, London.

Pacione, M. (1984a) *Rural Geography*, Paul Chapman, London.

Pacione, M. (1984b) The definition and measurement of the quality of life, in M. Pacione and G. Gordon (eds.) *Quality of Life and Human Welfare*, Geo Books, Norwich.

Pahl, R. E. (1965) *Urbs in Rure*, Weidenfeld & Nicolson, London.

Pahl, R. E. (1966) The social objectives of village planning, *Official Architecture and Planning*, Vol. 29, pp. 1146–50.

Pahl, R. E. (1968) The rural-urban continuum, in R. E. Pahl (ed.) *Readings in Urban Sociology*, Pergamon Press, Oxford.

Parker, K. (1984) *A Tale of Two Villages: The Story of the Integrated Rural Development Experiment in the Peak District, 1981–84*, Peak Park Joint Planning Board, Bakewell.

Pennant, T. (1774) *A Tour in Scotland and Voyage to the Hebrides 1772* (2 vols.), John Monk, Chester.

Perry, R., Dean, K. and Brown, B. (1986) *Counterurbanization: International Case Studies of Socio-Economic Change in Rural Areas*, Geo Books, Norwich.

Phillips, D. and Williams, A. (1982) *Rural Housing and the Public Sector*, Gower, Aldershot.

Phillips, D. and Williams, A. (1984) *Rural Britain: A Social Geography*, Blackwell, Oxford.

Philpott, J. C. and Tyler, G. J. (1987) Interpersonal variation in farm workers' earnings: analysis of Wages and Employment Enquiry data, *Journal of Agricultural Economics*, Vol. 38, pp. 163 72.

Popp, H. (1976) The residential location decision process, *T.V.E.S.G.*, Vol. 67, pp. 300–5.

Prattis, J. I. (1977) *Economic Structures in the Highlands of Scotland* (Speculative Paper no. 7), Fraser of Allander Institute, Glasgow.

Randolph, W. and Robert, S. (1981) Population redistribution in Great Britain, 1971–81, *Town and Country Planning*, Vol. 50, pp. 227–31.

Rennie, F. W. (1986) A community view of 'integrated rural development', in P. Selmen and L. Houston (eds.) *Environmental Conservation and Development*, Planning Exchange, Glasgow, Occasional Paper 24, pp. 94–104.

Rhind, D. W. (1983) (ed.) *A Census Users' Handbook*, Methuen, London.

Richmond, P. R. (1987) The social implications of State housing provision in rural areas, in D. G. Lockhart and B. Ilbery (eds.) *The Future of the British Rural Landscape*, Geo Books, Norwich.

RICS (1990) *A Place to Live: Housing in the Rural Community*, London.

Robinson, D. (1988) The changing labour market. Growth of part-time employment and labour market segmentation in Britain, in S. Walby (ed.) *Gender Segregation at Work*, Open University Press, Milton Keynes.

Robinson, G. M. (1990) *Conflict and Change in the Countryside*, Belhaven, London.

Robson, N., Gasson, R. and Hill, B. (1987) Part-time farming – implications for farm family income, *Journal of Agricultural Economics*, Vol. 38, pp. 167–91.

Rogers, A. (1987) Voluntarism, self-help and rural community development: some current approaches, *Journal of Rural Studies*, Vol. 3, pp. 353–60.

Rogers, A. and Winter, M. (1988) *Who Can Afford to Live in the Countryside?* (Occasional Paper no. 2), Centre for Rural Studies, Cirencester.

Roseman, C. (1971) Migration as a spatial and temporal process, *Annals of the Association of American Geographers*, Vol. 61, pp. 589–98.

Ross, D. and Usher, P. (1986) *From the Roots Up: Economic Development as if Community Mattered*, Lorimer, Toronto.

Rossi, P. (1980) *Why Families Move*, Sage, London.

Rowles, G. D. (1986) The geography of ageing and the aged: toward an integrated perspective, *Progress in Human Geography*, Vol. 10, no. 4, pp. 511–39.

Rugman, A. and Green, M. (1979) Demographic and social change, in F. Joyce (ed.) *Metropolitan Development and Change. The West Midlands: A Policy Review*, Saxon House, Farnborough.

Salveson, P. (1989) *British Rail: The Radical Alternative to Privatisation*, Centre for

Local Economic Strategies, Manchester.

Saunders, P. and Harris, C. (1990) Privatization and the consumer, *Sociology*, Vol. 24, no. 1, pp. 57–75.

Savage, I. P. (1985) *The Deregulation of Bus Services*, Gower, Aldershot.

Saville, J. (1957) *Rural Depopulation in England and Wales 1851–1951*, Routledge & Kegan Paul, London.

Scitovsky, T. (1987) Growth in the affluent society, *Lloyds Bank Review*, no. 163, pp. 1–14.

Scola, P. M. (1961) Scotland's farms and farmers, *Scottish Agricultural Economics*, Vol. 11, pp. 59–62.

Shaw, G. and Williams, A. (1985) The role of industrial estates in peripheral rural areas: the Cornish experience 1973–1981, in M. J. Healey and B. W. Ilbery (eds.), op. cit.

Shaw, J. M. (ed.) (1979) *Rural Deprivation and Planning*, Geo Abstracts, Norwich.

Shaw, J. M. and Stockford, D. (1979) The role of statutory agencies in rural areas: planning and social services, in J. M. Shaw (ed.), op. cit.

Shawyer, A. J. (1990) Farm structure and farm families: a Nottinghamshire field area. *East Midland Geographer*, Vol. 13, pp. 1–18.

Shucksmith, M. (1981) *No Homes for Locals?*, Gower, Aldershot.

Shucksmith, M. (1987) *Public Intervention in Rural Housing Markets* (PhD thesis), University of Newcastle upon Tyne.

Shucksmith, M. (1988) Crofter housing policies in Scotland, in R. Byron (ed.), op. cit.

Shucksmith, M. (1990) *Housebuilding in Britain's Countryside*. Routledge, London.

Shucksmith, M. and Lloyd, G. (1982) The Highlands and Island Development Board regional policy and the Invergordon closure, *National Westminster Bank Quarterly Review*, May, pp. 14–24.

Shucksmith, M. and Watkins, L. (1988) The supply side of rural housing markets, in A. Rogers and M. Winter (eds.), op. cit.

Shucksmith, M., Bryden, J., Rosenthall, P., Short, C. and Winter, M. (1989) Pluriactivity, farm structures and rural change, *Journal of Agricultural Economics*, Vol. 40, pp. 345–60.

Smith, J. A. and Gant, R. L. (1982) The elderly's travel in the Cotswolds, in A. M. Warnes (ed.), op. cit. pp. 323–6.

Smout, T. C. (1972) *A History of the Scottish People 1560–1830*, Fontana, London.

Somerset County Council (1986) *West Somerset Rural Development Area Programme 1986–89*, Taunton.

South Lakeland District Council (1980) *HIP Strategy Statement, 1981/82*, Kendal.

Speare, A. (1970) Home ownership, life-cycle stages, and residential mobility, *Demography*, Vol. 7, pp. 449–58.

Spence, N., Gillespie, A., Goddard, J., Kennett, S., Pinch, S. and Williams, A. (1982) *British Cities: An Analysis of Urban Change*, Pergamon Press, Oxford.

Staffordshire County Council (1978) *Staffordshire Planning Survey*, Stafford.

Stamp, L. D. (1949) The planning of land use, *The Advancement of Science*, Vol. 6, pp. 224–33.

Standing Conference on Rural Community Councils (1978) *The Decline of Rural Services*, National Council of Social Service, London.

Strathern, M. (1981) *Kinship at the Core*, Cambridge University Press.

Suffolk County Council (1987) *Rural Development Programme 1986–87*, Ipswich.

Sulaiman, S. (1988) Village 'stock taking', *County Council Gazette*, Vol. 81, pp. 76–7.

Summers, G. T. (1986) Rural community development (USA), in R. H. Turner and J. F. Short jr (eds.) *Annual Review of Sociology*, Vol. 12, pp. 347–71.

Swansea City Council (1978) *Gower District Plan: Report of Survey*, Swansea.

Swansea City Council (1989) *Swansea Local Plan*, Swansea.

Swanson, L., Luloff, A. and Worland, R. (1979) Factors influencing willingness to move, *Rural Sociology*, Vol. 44, no. 4, pp. 719–35.

Symes, D. and Marsden, T. (1983) Complementary roles and asymetrical lives: farmers' wives in a large farm environment, *Sociologia Ruralis*, Vol. 23, pp. 229–41.

Symes, D. G. and Marsden, T. K. (1984) Landownership and farm organisation: evolution and change in capitalist agriculture, *International Journal of Urban and Regional Research*, Vol. 8, pp. 388–401.

Symes, D. and Marsden, T. (1985) Industrialisation of agriculture: intensive livestock farming in Humberside, in M. J. Healey and B. W. Ilbery (eds.), op. cit.

Taylor, S. (1988) The elderly as members of society: an examination of social differences in an elderly population, in N. Wells and C. Freer (eds.), op. cit.

Thomas, F. G. (1939) *The Changing Village*, Nelson, London.

Thomas, R. and Elias, P. (1989) Development of the standard occupation classification, *Population Trends*, Vol. 55, pp. 16–21.

Thompson, J. (1987) Ageing of the population: contemporary trends and issues, *Population Trends*, Vol. 50, pp. 18–22.

Thorns, D. (1968) The changing system of rural stratification, *Sociological Review*, Vol. 8, pp. 161–78.

Thrift, N. (1987) The geography of late twentieth century class formation, in N. Thrift and P. Williams (eds.) *Class and Space*, Routledge & Kegan Paul, London.

The Times (1990) 18 August.

Transport and Road Research Laboratory, 18 August. (1980) *The Rural Transport experiments: Proceedings of a Symposium Held at the Transport and Road Research Laboratory, Crowthorne on 8th November 1979* (TRRL Supplementary Report 584), Crowthorne.

Tricker, M. and Martin, S. (1985) Rural development programmes: the way forward, in M. J. Healey and B. W. Ilbery (eds.), op. cit.

Town and Country Planning Association (1989) *The Future Planning of the Countryside: A Discussion Paper*, London.

Townsend, A. R. (1986) Spatial aspects of the growth of part-time employment in Britain, *Regional Studies*, Vol. 20, no. 4, pp. 313–30.

Townsend, A. R. (1987) Regional policy, in W. F. Lever (ed.) *Industrial Change*, Longman, London.

Townsend, P. (1979) *Poverty in the United Kingdom*, Penguin Books, Harmondsworth.

Turnock, D. (1969) Regional development in the crofting counties, *Transactions of the Institute of British Geographers*, Vol. 48, pp. 189–204.

Turnock, D. (1970) *Patterns of Highland Development*, Macmillan, London.

Tyson, W. J. (1988) *A Review of the First Year of Bus Deregulation*, Association of Metropolitan Authorities, London.

Vartiainen, P. (1989) Counterurbanisation: a challenge for socio-theoretical geography, *Journal of Rural Studies*, Vol. 5, pp. 217–25.

Veljanovski, C. (1987) *Selling the State: Privatisation in Britain*, Weidenfeld & Nicolson, London.

Vining, D. R. and Strauss, A. (1977) A demonstration that the current deconcentration of population in the United States is a clean break with the past, *Environment and Planning A*, Vol. 9, pp. 751–8.

Wagstaff, H. R. (1970) Scotland's farm occupiers, *Scottish Agricultural Economics*, Vol. 21, pp. 277–85.

Wagstaff, H. R. (1971) Recruitment and losses of farm workers, *Scottish Agricultural Economics*, Vol. 21, pp. 7–16.

Wagstaff, H. R. (1972) Recruitment, labour turnover and losses of full-time male workers in Scottish agriculture 1967–70, *Scottish Agricultural Economics*, Vol. 22, pp. 153–5.

Walford, N. S., Shearman, J. and Lane, M. (1989) *Rural Areas Database Catalogue: Data Directory and Guide to Services*, ESRC Data Archive, University of Essex.

Walker, C. and McCleery, A. (1987) Economic and social change in the Highlands and Islands, *Scottish Economic Bulletin*, Vol. 35, pp. 8–20.

Warnes, A. M. (ed.) (1982) *Geographical Perspectives on the Elderly*, Wiley, Chichester.

Warnes, A. M. (1987) Geographical locations and social relationships among the elderly, in M. Pacione (ed.) *Social Geography: Progress and Prospect*, Croom Helm, London.

Warnes, A. M. and Law, C. M. (1984) The elderly population of Great Britain: locational trends and policy implications, *Transactions Institute of British Geographers* (New Series), Vol. 9, no. 1, pp. 37–59.

Watkins, C. (ed.) (1989) *Teleworking and Telecottages* (Occasional Paper no. 4), Centre for Rural Studies, Cirencester.

Watkins, C., Blacksell, M. and Economides, K. (1988) The distribution of solicitors in England and Wales, *Transactions, Institute of British Geographers* (New Series), Vol. 13, pp. 39–56.

Watkins, C. and Winter, M. (1988) *Superb Conversions?*, CPRE, London.

Watkins, R. (1979) Educational disadvantage in rural areas, in J. M. Shaw (ed.), op. cit.

Watson, W. (1964) Social mobility and social class in industrial communities, in J. Gluckman (ed.) *Closed Systems and Open Minds*, Oliver & Boyd, London.

Webber, R. and Craig, J. (1977) *A Socio-Economic Classification of Local Authorities in Great Britain* (OPCS Studies in Medical and Population Subjects), HMSO, London.

Weekley, I. (1988) Rural depopulation and counterurbanization: a paradox, *Area*, Vol. 20, pp. 127–34.

Wegner, G. C. (1982) Ageing in rural communities: family contacts and community integration, *Ageing and Society*, Vol. 2, pp. 211–29.

Wegner, G. C. (1988) *The Supportive Network*, Allen & Unwin, London.

Welch, R. (1984) The meaning of development, *New Zealand Journal of Geography*, Vol. 76, pp. 2–4.

Wells, N. and Freer, C. (eds.) (1988) *The Ageing Population – Burden or challenge?*, Macmillan, London.

Wenger, C. (1982) The problem of perspective in development policy, *Sociologia Ruralis*, Vol. 12, pp. 5–16.

West Glamorgan County Council (1980) *County Structure Plan*, Swansea.

West Midlands Planning Authorities Conference (1979) *West Midlands Regional Strategy, Report of the Joint Monitoring Steering Group*, Birmingham.

Whatmore, S. (1988a) From women's role to gender relations. Developing perspectives in the analysis of farm women, *Sociologia Ruralis*, Vol. 28, no. 4, pp. 239–47.

Whatmore, S. (1988b) *The Other Half of the Family Farm: An Analysis of the Position of Farmers' Wives in the Familial Gender Division of Labour on the Farm* (unpublished PhD thesis), University of London.

Williams, A. (1973) *The Highlands and Islands Development Board 1965–70: Policy-Making in an Administrative Setting* (unpublished M Litt thesis), University of Glasgow.

Williams, A., Shaw, G. and Greenwood, J. (1989) From tourist to tourism entrepreneur, from consumption to production: evidence from Cornwall, England, *Environment and Planning A*, Vol. 21, pp. 1639–53.

Williams, G. (1984) Development agencies and the promotion of rural community development, *Countryside Planning Yearbook*, Vol. 5, pp. 62–86.

Williams, G. (1985) The achievement of specialist agencies in rural development, in S. Barrett and P. Healey (eds.) *Land Policy: Problems and Alternatives*, Gower, Aldershot.

Williams, R. (1975) *The Country and the City*, Chatto & Windus, London.

Williams, W. M. (1963) *A West Country Village, Ashworthy*, Routledge & Kegan Paul, London.

Winyard, S. (1982) *Cold Comfort Farm: A Study of Farmworkers and Low Pay*, Low Pay Unit, London.

Wolpert, J. (1966) Migration as an adjustment to environmental stress, *Journal of Social Issues*, Vol. 22, pp. 92–102.

Wrekin District Council (1989) *Wrekin Trends*, Telford, Shropshire.

Wright, S. (1983) Pigeon-holed policies. Agriculture, employment and industrial development in a Lincolnshire case study, *Sociologia Ruralis*, Vol. 23, pp. 242–60.

Yuill, D. (ed.) (1982) *Regional Development Agencies in Europe*, Gower, Aldershot.

INDEX